First they ignore you.
Then they laugh at you.
Then they fight you.
Then you win.
—Mahatma Gandhi

BROKEN WING

BIRDS, BLADES, AND BROKEN PROMISES

JOHN H. GRAVES, ChFC, CLU

BROKEN WING
BIRDS, BLADES, AND BROKEN PROMISES
by John Graves, CLU, ChFC

Safe Harbor International Publishing
121 N. Fir St.
Suite C
Ventura, CA 93001

This publication is designed to provide accurate and authoritative information in regard to the subject matter covered. It is sold with the understanding that neither the author nor the publisher is engaged in rendering legal, accounting, securities trading or other professional services. If legal advice or other expert assistance is required, the services of a competent professional person should be sought.

ISBN: 978-0-9835731-4-2

Cover and interior design
by GKS Creative, gkscreative.com

Printed in the United States of America

Dedicated to...
Jesse Grantham, a brilliant birder
Leif Anderson, a brilliant winder

"Over increasingly large areas of the United States, spring now comes unheralded by the return of the birds, and the early mornings are strangely silent where once they were filled with the beauty of bird song."

RACHEL CARSON, *Silent Spring.*

TABLE OF CONTENTS

PREFACE

http://www.youtube.com/watch?v=1RcTjdY1aN4

He soars, searching for carrion. With a new chick in the nest and the mother in constant attendance, he is busy today, every day, searching, seeking sustenance. They have just a few months to bring the youth to the edge of the aerie, to watch her try her new wings, to see her fly. Her own search would then begin. Their race would survive another winter, another spring.

Whoomp.

He circled the tiny, dark image of death below. Food. His action would bring others of their tribe. A kettle forms in the sky as, one by one, they spy a fellow Geier circling a potential meal. They prowl the skies seeking food from death. Their natural job is to clean the hillsides of dead animals. They eat the remains of what others have killed. Or of what has died unnaturally.

Whoomp. Whoomp.

Like Lucifer stuck below heaven, the wind turbine's white wings strum the air. Their slow movement gives an illusion of grandeur, of greatness. Plugged into the ground with their deep seated leg, they too seek sustenance, from the air. They seek to draw power from nothing, from the wind. Like the Avatar's ground eating machine, they consume everything in their path.

Giannis glides gracefully across the blue white morning sky. At times he thinks the griffon is playing in and out of the blades, as if in a dance. How soon it would become danse macabre… He is a paraglider with a GoPro HD cam, enjoying a cool morning thermal. The lift is subtle, the first of many to warm the island of Crete. From 600 meters he can see the wine dark sea surrounding his aery vigil. The griffon can see the dark smudges moving towards him, others from the tribe seeking sustenance. The paraglider could also see—down into the powerful blades.

Whoomp. Whoomp.

He zooms in on the majestic carrion eater. As you watch his eyes, he sees only the meal, only the fodder for his mate, their chicks. For agonizing minutes you watch his slow gyration across the heavens, descending to dinner.

The blade slices his wing, nearly severing it from his body. His graceful gyre becomes a death spiral. He falls gracelessly to Earth.

Giannis lands by the majestic bird. The camera records his staggering to get to his feet. He struggles, not understanding. His wing lies useless. He staggers again, collapses. The camera zooms in. His last moment? His eyes are searching for—his meal, their nest, the future.

Is all lost? Will he become the meal for another? Will his mate have to abandon the nest in search of their meals now? His chicks may fall to another bird of prey, perhaps a golden eagle or another griffon. Her life, her line, may end with this Avatar, this foreboding of the future.

The death of birds at the hands of these 400' Franken Towers is wide spread—and growing. The story here is from England.

British birders—twitchers—flocked to the Outer Hebrides, the northernmost islands of the British Isles, in the summer of 2013 to see the rarest of birds, the white-throated needletail. Dozens were pleased with sitings all morning, only to watch in horror as a local turbine sliced it to death. Not since 1991 had the genus been sighted there—and not soon will it return.[1]

The following video is a longer, more detailed update of a wide number of kills across Europe. This is very unpleasant to watch, so be warned.

https://www.youtube.com/watch?t=3&v=S-ppZhISYd8

The raptor genus is under attack. The African species has been downgraded from threatened to endangered. The European griffon is critically endangered, the next level down. Only extinction remains. Mankind's agricultural encroachments, captures for animist fetishists or medical investigators, and electrocution are the main threats.

Today, wind turbines present the final threat to the genus. Across the Earth's Northern Hemisphere, griffons, vultures, osprey, hawks, eagles, falcon, and owls are each threatened by these new Avatars of the air. Wind turbine farms are stretched across the very flyways in the sky that these birds use in migration and habitation. The turbines seek what the birds seek—wind. The turbine farms are growing in size, in population, and in killing power. Today these tax farms have a license to kill.

As the number of wind farms increase, the *take* increases. This is the government's euphemism for a "license to kill." No James Bond heroes here, folks. This is a slaughter most foul. Six million birds are killed each year in Spain, where the turbine numbers are greatest. 537,000 is the government's estimate from the U.S. Department of Fish and Wildlife for 2012. 83,000 were raptors, the hunters of the skies. 888,000 bats die every year.

No actual figures are released. You have to take their word on this. Firms are not required to publish their takes. When they do, the figures are entirely unscientific and unreliable. The Administration refuses to release the data behind their figures, saying it would "expose trade secrets or implicate ongoing enforcement investigations."

There are no investigations. There is no enforcement. Just the opposite. All wind farms have, on December 9, 2013, been given a five year right to kill.[2] This has recently been increased to 30 years. Why? To ensure that the operators and their stakeholders have assurance of risk-free returns.

In 2009, PacifiCorp was fined $10.5 million for their alleged responsibility in the deaths of 232 eagles on power lines. Was it because this is a coal fired power plant company, one that had to be taught a lesson? The 20 eagles found dead at the same company's wind farms have resulted in neither investigation, prosecution, nor fines. Duke was recently fined $1 million for the deaths of birds at their Wyoming farm, a first in the U.S. wind industry. Electrocution merits massive fines—severing bird wings is apparently acceptable because it is saving the planet from the scourge of humanity.

The absence of prosecution is an assurance that many more birds will die. By 2030, the U.S. government figures are estimated at 1.4 million wind farm avian deaths each year. Globally, the estimate is 4 million.[3] The impact is felt greatest at the top of the food chain. Raptors have smaller populations. They breed for longer periods and their young remain nest bound longer than others.

The sheer volume of bird kill does not begin to depict the magnitude of ecological damage, since the most susceptible species tend to be those which are *keystone species.*

Proponents of large-scale wind turbine sites (including some federal agencies) tend to favor sparsely vegetated saddles or other funnel like landforms, which are highly correlated with high density bird migration routes or raptor soaring locations.[4]

The source of this quote? *The Encyclopedia of Earth,* hardly a right wing extremist group. "The EoE is a free, expert-reviewed collection of environmental-based content contributed by scholars, professionals, educators, practitioners, and other experts who collaborate and review each other's work." This is not the junk science storefront of Wikipedia, where any and all may write as they wish, subject to the renowned internal bias from the Board.[5] This is expert-reviewed material, as free of slant as is possible today.

The word is out. Site location for 30-story wind turbines in migration alleys is encouraged. Stable wind sourcing from a common direction for long hours is the better choice for siting towers that generate the power. Report your takes, or not, as you like. If they exceed the permitted numbers, apply for a variance. No, you don't need to spend more than the federally mandated cap on take mitigation. No needs for a full Environmental Impact Report (EIR). Just keep building. *Gotta put those coal plants outta business.*

Many of the raptors eviscerated by these manmade killing machines are protected under federal environmental laws. Many eagles are endangered species. Five species are approaching extinction here and globally: the golden eagle, sea eagles, bustards, whooping cranes (not a raptor, but endangered and under attack in Europe and the Midwest) and the Tasmanian eagle. The golden eagle population in the western U.S. was in such decline in 2009 that the feds made it a policy to prevent even a single death. Too bad for the birds. Hundreds have died since then in the blades of wind farms.[6] An unknown number of eagle families have been destroyed by the deaths of the bread winner. By some estimates, there are fewer than 500 golden eagles left in California, once the ancestral home to thousands of families. Wind turbine farms cover hundreds of acres. They are often built on ridges along migratory pathways—for raptors and their prey.

Annual avian deaths are estimated (via meta analysis, or the review of all published material) between 140,000 and 328,000. Little access to random-based studies is available. These studies are nonscientific. Most feel these studies probably under-weight the count, according to authors of the study of 10/2013.[7] If the current Department of Energy (DOE) requirements that 20% of U.S. energy production come from *alt-en* sources by 2030, an estimated 1.4 million avian deaths per year is on the low side of guesses.

The authors of this study were frustrated by the lack of publicly-available data on bird collisions in many regions of the U.S., including the entire southwest. They acknowledge that this may have skewed their mortality estimates somewhat, but the only way that such models can be improved in the future is if industry reports no longer remain confidential.[8]

Raptors command the top of the food chain. They are apex predators. Their range is significant. Their numbers less so. Their reproductive rates are low while their life spans are long. They have no known natural enemy other than Man. The following raptors are among those found most commonly at wind farm sites globally:

- Griffin falcon *(Gyps fulvus)*
- Golden eagle *(Aquila chrysaetos)*
- Bald eagle *(Haliaeetus leucocephalus)*
- Old world kestrel *(Falco tinnuculus)*
- American kestrel *(Falco sparvenius)*
- Brown pelican *(Pelecanus occidentalis)*
- Northern gannet *(Morus bassanus)*

These populations are rapidly becoming frail. Survival is a challenge at any time. It is nearly impossible today. Tomorrow will be too late. They will die.

Consider the California condor. Takes are now allowed against this raptor in California's Tehachapi Mountains. This, after millions have been spent over two decades to protect these amazing wind walkers. The whooping crane, also near extinction and legally protected, has fewer than 300 members. It is now subject to take in the Midwest. Both examples suggest selective removal of apex and endangered species. Wind turbine industrialists are not maliciously seeking out birds for destruction. The deaths result from poor understanding of avian habits, a lack of awareness of the scope of the rising challenge, and a desire for assured profits. All of these are kept in the air with the climate change fear of a politicized world.

Raptors may suffer extinction in a decade but their species is not alone in this slow death spiral. A 2003 study[9] found that 78% of *passerine* deaths from wind turbines were of endangered species as defined by the U.S. Migratory Bird Treaty Reform Act. The authors of the study question the veracity of their own claims, as the paucity of evidence skewed their findings. They doubted the figures were representative of actual deaths.

Why? Another report, the infamous APWRA of 2004[10] says:

We found one raptor carcass buried under rocks and another stuffed in a ground squirrel burrow. One operator neglected to inform us when a golden eagle was removed as part of the WRRS. Based on these experiences, it is possible that we missed other carcasses that were removed.

Why would an operator remove or hide carcasses? Reporting is costly and adversarial to the regime. It is essentially unnecessary today. Let's just move on. We can do without these details.

An annual compounding of deaths for any species leads to a species-wide death spiral. It is happening to fish populations in all oceans. It is happening to large carnivore populations across Africa and Asia. Those at the top cannot afford this cycle of death. Now we are killing the top of the avian food chain. The bald eagle, our symbol of American might, is today a symbol of our politicians and capitalists ignorant righteousness.

What is the response from the wind energy industry to these charges? Industry and government comments go first to relativistic morality. Here are a few of their remarks:

Hundreds of millions of birds are killed by cats, cars, buildings, and one another. The deaths of a few hundred thousand each year pales by comparison. We kill now so that future generations of birds may live.

They lump all birds into one category, then belittle the deaths of a small number. This is statistical elitism. Cats are more dangerous to falcons, eagles, and vultures than wind towers. Really?

Their comments trend immediately to Climate Change, human induced. Guilt rides to the forefront. CO_2 emissions must be reduced or we are all going to die, all species. Wind is free and easy: to harness, to extract and to divine.

These tradeoffs are the consequence of concerns about AGW, anthropogenic global warming. This agenda have risen to the top of many political arguments. While I have views on the subject, I shall leave it to a far more intelligent scientist to express his perspective from 91 years of applied research.

Freeman Dyson replied to an interviewer for *The Register* on 10/11/15:[11]

Are climate models getting better? You wrote how they have the most awful fudges, and they only really impress people who don't know about them.

I would say the opposite. What has happened in the past 10 years is that the discrepancies between what's observed and what's predicted have become much stronger. It's clear now the models are wrong.

For the past eight years, a billion dollars each year has been given of your tax dollars to encourage alternative energy development: alt-en. Rentiers have risen on the updraft of this massive capital supply. An avian slaughter has ensued, in the name of protecting the environment. The rationale? 50% or more of all birds will die from global warming by 2100. Better to allow a few to die now, to protect the entire populations of all birds, indeed of all species.

Must we kill these raptors to protect them from our selfish actions? Or, are our selfish actions killing these birds?

To explore this challenge of the threat of future extinction, we need to step back. We need to have a strong understanding of…

- how wind works,
- how turbines generate power,
- how various nations are developing this power,
- how successful it is at its stated quest – CO2 reduction
- each of the many threats wind presents to man and wildlife
- state of the art developments evolving in wind technology
- what you can do

This book will take five positions:

1. Wind turbines are an ancillary source of power generation.
2. Avian deaths are unacceptable results of wind turbine growth.
3. Alternative technology solutions are abundant and available.
4. Government mandates distort the wind market.
5. Human illness and death are direct results of wind energy.

You will learn positive, constructive approaches to alternative energy sourcing. I shall urge these four rather straightforward ideas on my readers:

- Wise wind power manufacturing, siting, and sourcing
- Far greater awareness and prevention of avian mortality
- Turbine and distribution design choices in an open marketplace unfettered by regulatory or preferential intervention
- Elimination of government price supports for production, siting, and distribution.

The rapid increase in wind turbine technology has allowed more and more efficient power production over the recent past. This may be encouraged, but not at the expense of large numbers of our avian fellows. As the number of wind turbines increase, their siting, design, and production encourage more avian deaths. These deaths can be avoided, reduced, or completely eliminated with a wider variety of turbines.

Government mandates, put in place by a *rent-seeking* industry, forced the use of tall spindled, three-blade monopole devices. While more energy efficient than many previous designs, a wide array of designs are available today that reduce or eliminate avian casualties. These choices are far more efficient and efficacious power sources. These designs are applicable in a wide array of situations. The monopole spindle may be right for a limited number of sites. It is the wrong choice for most sites. Adopting the tool to the site should make sense. Killing to adopt is simply wrong. Today, it is one size fits all.

We will study these fields of wind energy research:

- Physics of wind
- Wind energy, its production and distribution
- Economics of wind power
- Environment and wind energy
- Wind energy and its effect upon avian and human health

We'll argue for a lowering of federal regulatory interference in a now mature industry. Allow a thousand turbine designs to shine. Spend time and resources on electrical power storage choices, a most difficult engineering feat.

The market for wind technology is strong and growing across the globe: $6.9 billion in 2013. Remove the regulatory constraints and preferential tax treatment of their implementation and follow-on energy sales. The invisible hand of the market is far better able to determine best practices in the wind turbine and power distribution market than top down mandates from an industry dominated by its own regulators and lobbyist groups.

Those mandates are designed by an industry addicted to government funds—indeed, dedicated to these tax dollars. This capital is the more egregious because of the source. Any industry that is so coddled by a government that it can write its own laws and regulations has removed itself beyond the pale. No industry should be omniscient regarding its future. Or ours.

Finally, make no mistake. Eagles, hawks, owls, and all raptors are no more friendly or cuddly than a polar bear. These creatures are vicious, savage hunters. They live by their next kill. They are unafraid of any beast.

http://www.youtube.com/watch?v=uPUP8ey3MjE

Be warned. Do not watch the above video if you are squeamish. You will see eagles kill an owl, a deer, and a wolf. The dear and wolf are several times the size of the eagles. The Kazak eagle hunters run their birds in packs. I once held a golden eagle wolf hunter on my arm in the Tien Shen mountains of Kazakhstan, wondering which eye he would pluck out first as he stared at me so closely, so intent. No one believed my outrageous stories of wolf-killing eagles until I shared these videos.

There is nothing nice about the bald eagle or the golden eagle. They are natural born killers. This book is not meant to prey on your fancy urban concerns for poor helpless wildlife. These masters of their universe can take care of themselves—usually. Even in captivity they are dangerous. These are not pets.

I have seen raptors in the wild on five continents. I have visited several Raptor Centers in the U.S. (www.orc.org). These birds can and do can attack without warning and are not friendly to humans. If you want cuddly, buy a stuffed animal.

This book is about their useless slaughter at the hands of an even more rapacious beast—Man. This slaughter is unnecessary, even by the irredentist standards of these capitalist rent seekers. No good comes from these deaths.

Why have I written this book?

First, the story needs to be told. Some people know of a few of these issues. Some know of all of these problems. The media certainly isn't doing their investigative journalism very well. If you know that

- birds and humans die because of wind turbines,
- CO2 emissions increase with wind turbines,
- electricity costs spiral from price supports and tariffs,
- your tax dollars are wasted on these capital boondoggles, and
- other choices are readily available,

wouldn't you want to be aware of other choices? Wouldn't you want to ask more questions? Wouldn't you demand serious answers to important scientific questions about avian mortality, human health, and expensive tax-supported projects?

Second, this book acts as a collection point for all the information currently available on the negative aspects of wind energy technology. Search the bookshelves. There are a few books on wind energy technology, a host of books on climate change, a very few on the human health aspects of wind power sourcing and even fewer on alternative wind energy choices. I am attempting to compile from a wide variety of global resources as much information as possible. The story constantly changes. The change works against the industry. They tell you their story. Learn more here.

Third, this is not an academic study of the wind industry. Such a study is virtually impossible, as the industry keeps its secrets close. It refuses to allow outside studies of any aspect of the wind story. Double blind scientific studies of the mining of the basic materials for wind turbines, of the siting and construction and interface with avian populations, of human health issues resulting from home proximity—each of these is forbidden. The federal governments of several nations are complicit in this. Each accepts the industry's expectation that any information, any at all, is confidential corporate information. Even simple access to the industry's annual reports is costly ($550.00) unless you are a member of the elite.

This book is written in the terms and attitudes of Sinclair Lewis and Rachel Carson. I realize I cannot hold myself as their equal. I can hope that the book holds itself as a light in the darkness, much as their books revealed what others wished hidden.

If this book succeeds, it will not be because of the author. It will be successful because of the readers. You will decide if the story is worthy of telling. Is it worthy of repeating? I cannot change a multi-billion dollar industry supported by the infinite resources of many federal governments. My views on the issues here are far less important than your views, your understanding.

I am not trying to change or even impact your views on climate change or economics or energy. I am interested in your views on *capitalist in-roaders* stealing your tax dollars to further their own interests, while mouthing exiguous support for environmental doggerel.

My view is: **If it ain't right for the petroleum industry, it ain't right for the wind industry.**

If after careful reading and thought you feel your purchase is a waste of money and time, so be it. Nothing ventured, nothing gained. If after careful reading and thought you feel your purchase is of value, then take the next steps. Share this book with others. Join a bird group. Urge your politicians to review the facts. Vote your preferences.

Birds die.

People sicken.

You can change this!

INTRODUCTION

Today's wind energy industry is a disaster. It is incestuous. It is owned by and simultaneously owns big government. Regulatory fiat defines business success. Rent seeking is the ideal business model. A few firms dictate to the regulators while taxpayers foot the bill.

This is an aberration of the truth. Birds die, billions are wasted, and people sicken and die on our farms and in the faraway deserts of Inner Mongolia. Few consumers and fewer environmentalists are aware.

The grid will fail, soon. One of the causes will be intermittency of power supply during peak demand time. The other cause will be the regulatory *force majeure* of closing hundreds of coal and gas fired power plants.

Wind turbines cause significant avian mortality across a wide range of bird genera. Many of these dying birds are endangered species. They are supposed to be protected under federal law. The accepted mortality figures from USFWS are guesstimates. They are absurdly low. They are based upon formulae rather than carcass counts. Siting issues are entirely ignored in the design process. Towers are placed in direct line with bird flyways. Alternative placement is never considered. Kills are relegated to the end of the line: "We'll get to them eventually. We have to save the planet right now."

Wind is, at best, an ancillary source of power generation for the nation. Intermittency is the largest engineering obstacle. From this challenge, most other issues arise. Face it and solve it. Smaller challenges will then be more accessible to solution.

The industry is entirely supported by the governments of the world. These governments fear falling skies and rising waters. They proffer tax incentives for the construction, maintenance, and sale of power from wind turbines. They remove all risk from the industry. The result is a fattened calf. The result is massive tax fraud perpetrated by the governments of the world upon their own peoples.

Wind turbines are meant to reduce CO_2 emissions for the industrialized world. They are meant to replace coal and gas-fired power plants. These are falsehoods. Power plants need to cover for their power production intermittency. New turbines mean more coal and gas-fired power plants. These produce more CO2. They produce more of it than they normally would because they have to remain online even as the turbines are effective. Construction of turbine towers require massive amounts of steel, concrete, and hydrocarbon-based plastics.

Wind turbines are directly responsible for absolute and relative human health concerns in China. Cancer and industrial accidents take an unknown toll of human suffering in the wastelands of Inner Mongolia. These lands have been transformed from desert to waste via mining of rare earth minerals for wind turbine permanent magnets. These turbines are sited too close to farms and residences. The confluence of health issues as expressed by these human neighbors and turbine towers is currently coincidental. This is only because most medical researchers refuse to investigate. The suffering is documented.

Alternative choices abound: in siting, construction, avian mortality, design, and human health. Some of these choices are tested. Some are simply interesting ideas. Many are worthy of consideration. Sites can be better monitored prior to construction over a period of years. Tower choices can be expanded beyond the monopole three blade choice. Human habitation should be much further afield from turbine towers.

Our tale has many players. Protagonists become antagonists. Antagonists fall prey to good works and change their hair shirts. *In medias res* is the simple Greco-Roman means of telling a story, one that starts in the middle, then evolves to the beginning before finishing in tragedy. Good guys become bad guys. Bad guys change their colors. Both collude and collide. Over all, the *ludi magister* presides—the Master of the Games. He is The Federal Regulator.

We have much ground to cover.

We shall take our wind story in courses. Each course will be introduced by a raptor. Enjoy the visuals of each bird's life. While dangerous hunters, we can respect their place in the Domain of the Sky. These videos are intended to put you up front and personal with the birds. Know them from a safe distance. Appreciate their simple grandeur.

First we shall explore the emerging world of wind energy. What is the source? How does it work? How is it exploited, by others as well as by humans? How do we measure the impact of our use of it on society, on the wilds? Are we informed on the environmental consequences?

Then we shall discuss Aeolian technology and turbines—*les aeolian gigantes*, as the French call them. We shall look at wind energy production, availability, distribution, and use. The industry has grown enormously across Europe and America. What has driven this growth?

We shall then dip into my favorite, economics. How do the economics of supply, demand, and pricing affect the industry? How does legislative, regulatory, and legal action affect this mature industry? What are the impacts upon local society? What are the differing approaches taken in sample countries: Spain, Germany, Great Britain, and America?

Next let's look at the environment. What is the environmental impact of these new wind farms? We shall examine the estimated global climate impact, regional impacts, life cycle assessments, and human intervention techniques relative to mitigation issues. We shall attempt to discern both the positives and negatives. Of particular concern are two issues: effects upon avian populations and effects upon humans. The human health concern is *wind turbine syndrome.* The impact upon the miners who reach deep for neodymium—the heart of every wind turbine—is clear.

What are the alternative wind development tools? How does private capital innovation compare to government tax revenue regulation? What does today's wind turbine technology offer at the cutting edge? How can these new ideas replace older approaches? Are they cost effective at the economic and environmental levels? Can regulators move sufficiently quickly for adaptation? Are there effective avian mortality prevention devices for wind farms?

We shall summarize our discussions with three forms of stimulation. We shall stimulate personal action, local action, and national action.

This action call has come full circle in its demand from the people. Once, a majority of the electorate wanted a fair shake for all. Unions, Democrats, and other politicians led the fight against Big Capital. For years they won the battles. Then the Berlin Wall fell. Communism was defeated. Suddenly there were far fewer enemies to fight. For a variety of reasons, carbon became the enemy. Extraordinary how the very basis of life itself became the cancer that had to be destroyed. In the struggle against this new nemesis, many people found meaning. Legislation was drafted, meetings were held around the world (in very nice locales), and regulations followed the law.

Funny about regulations. They make a few people a lot of money. It's the power game: trick the crowd with games, write the right ideas into law, grow administrative application, and steal the gold from their hearts.

The top 20% of California's population receive 60% of the wealth transfers, also known as tax credits,[12] while the rest eat the rising dirt of the enforced draught. California has the highest poverty rate in America. How many poor Latinos earning $36,000 a year can afford a Tesla? How many care? How many can't even afford their rent and utilities? The Game is plagued against the poor. It always has been. Energy sourcing is simply the latest ruse, the carbonized Rube Goldberg variation.

Oddly enough, the wind turbine industry has increased carbon dioxide emissions. Europeans have led the field in alternative energy implementation, yet carbon dioxide emissions have increased substantially in Europe. For example, Denmark generates as much as 25% of its power requirements from wind, yet it has yet to close a single coal fired power plant. Why? Intermittency. The wind is variable. Flemming Nissen, head of power development at the Danish utility ELSAM, says, "Increased development of wind turbines does not reduce Danish CO_2 emissions."[13]

The electricity grid demands normalcy. *Shadow capacity* backs up the turbines with coal, nuclear, and gas-fired plants. These emit CO_2. More wind, more shadow, more emissions. It is a simple equation. It has nothing to do with politics or the environment. It is simply physics. Electricity demand drives the grid. The grid must be maintained at capacity. Intermittent electricity delivery does not maintain capacity. It must have back up. This is carbon intensive. The cycle rises with the number of turbines. These oddities plague the industry.

Broken promises ensue from the claims of regulators and politicians.

The claim to lower CO_2 emissions is a broken promise.
The claim to free energy is a broken promise.
The claim to positive environmental impact is a broken promise.
The claim to an efficient energy source is a broken promise.
The claim to more jobs is a broken promise.
The claim to lower electricity costs is a broken promise.
The claim to positive health results is a broken promise.
The claim to a greener planet is a broken promise.
The claim of no harm to birds and bats is a broken promise.

Facts are fungible. They can change. Today they can be made to say anything the speaker, or writer, wishes them to say. This is a shame. This is why there are a plethora of endnotes to this volume. Doubt what I write. Verify the source.

Compare it to other sources. Know these for yourself. Don't take my word for it. All writers have agendas, as do I. Seek beyond my agenda to the story. My opinion is of far less importance than the facts as you understand them!

Your judgment will determine the ultimate value of this tome. You will pay the price of admission and hopefully enjoy the reward of a good read. If you acquire the work in eBook format, feast on the videos. Be warned that some are very graphic—the avian videos in particular. Enjoy the graphics and cinematography. It is such a joy to design a book today with this level of user interface!

What is wind? How does it provide energy? What are the environmental consequences of wind turbines?

CHAPTER ONE
KITE

You must not fool yourself—and you are the easiest person to fool.
— **Richard Feynman**

Simply Scottish.com: Andrew MacDiarmid
Red kite from The RSPB on Vimeo:

https://player.vimeo.com/video/80364502

Among the most delicate of hunting birds, kites hover as buoyant as the air in which they fly. Their gliding and soaring is a graceful aerial ballet. Hanging motionless, the bird will roll, then rise vertically several hundred feet into the air, the ultimate airborne acrobat.

The god Isis often took the form of a kite in ancient Egyptian hieroglyphs. The angled wings and forked tail are distinctive in the sky. Insects, frogs, lizards, and small birds make for a varied diet. One species must be French because it only eats snails. Found across the globe except in arctic and extreme desert environments, they court in mid-air. Where else? Both parents build a nest of twigs and moss, home for their babies for the first six weeks of life.

Kites die by the dozens in wind turbine sites. Recently, a collision model has been developed that enables quantifying collision rates for red kites based on the distance between the wind turbine and the aerie of the birds. This model is based on physical characteristics of the wind turbines and behavioral parameters of the red kite such as flight height, flight frequency in relation to the distance to the aerie, and behavioral reactions to obstacles such as wind turbines.[14] See how effective the model is in reducing collision rates.

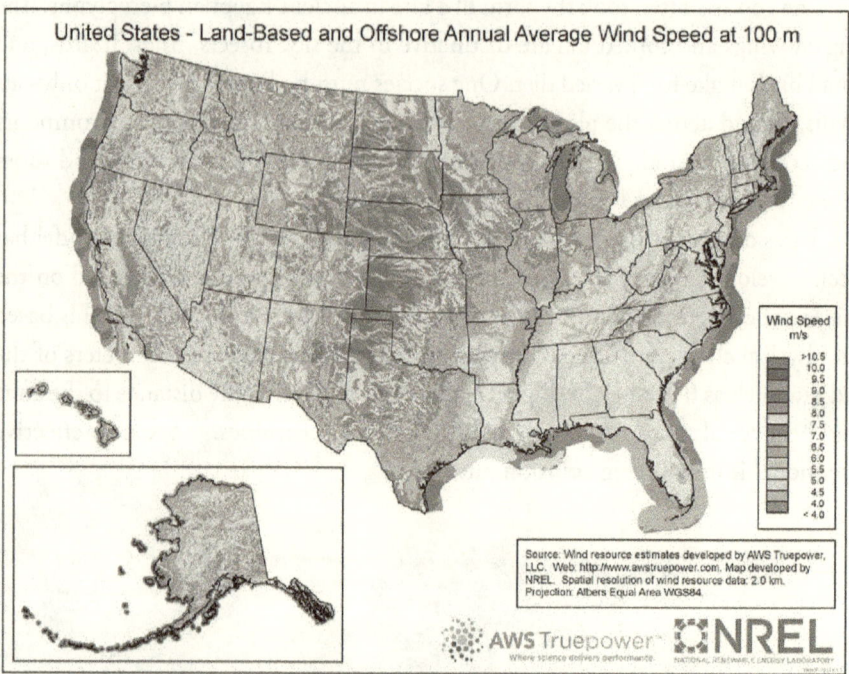

Current Wind Turbine Locations

http://eerscmap.usgs.gov/windfarm

Let's do a little science. Wind energy is much more than turbines spinning electricity for human consumption. In its many forms, wind energy is one of the *sub rosa* life supports on the planet. Wind results from a temperature (pressure) difference. Air moves from high pressure to low pressure. Simple. Every sailor, every pilot lives by the wind, the wind he manufactures with his sail, his wing. Sailors, pilots, and paragliders understand wind instinctively. One side of a wing or sail has higher pressure than the other side. Movement ensues, pulling the hull or plane towards this higher pressure.

In the atmosphere, the wind is the flow of energy from a high pressure system to a low pressure system. On a planetary scale, sunlight (solar radiation) heats each area according to its reflective and absorptive power. The rotation of the planet roils the resulting temperature (pressure) differences in the Coriolis Effect, forcing the wind to rotate—clockwise in the Southern Hemisphere, counterclockwise in the Northern Hemisphere. This rotation becomes fronts, storms, hurricanes, tornados, and cyclones. It also becomes high pressure systems—days of miraculous sunshine and clear skies.

Wind is also affected by turbulence. The ground can create this turbulence, as rivers, hills, seas, and mountains. Wind speed naturally increases with height above the ground because of *ground drag* (the friction caused by wind pressure against surface and near surface irregularities). Grass, rocks, trees, and buildings are a few sample irregularities. A wind speed of 20 mph at 1,000 feet reduces to 5 mph at 10 feet above a grassy surface. Different surfaces have differing effects. Turbulence flows from ground clutter—an abundance of interfering structures such as houses, trees, and garages. It creates swirling air streams that move upwards and downwards along the wind flow.

Mankind's cities affect wind, too. Stand next to a tall building and feel the air rush downward, across or uplift, stirring leaves and branches, skirts and bags. The local wind encounters the vertical surface and moves around it. In doing so, its speed increases—the pressure on the other side of the building may be lesser than on the windward side. A calm day with gentle breezes begets stronger wind currents around obstacles.

Rivers, lakes, and the sea affect the wind, driving it and dampening it. Water tends to absorb heat slower and retain it longer than earth. Thus, heated air from the land tends to blow offshore in the morning, while warmer air in the afternoon reverses the airflow and blows coastwise. You can *make your easting* for thousands of miles if you know your wind currents. European sailors traversed the Atlantic by *running down their easting*, then turning to starboard and heading for the Caribbean. In reverse, the Spanish galleons did the same as the traversed the Pacific from the Philippines to Panama. Chinese sailors did the same as they repeatedly crossed both the South China Sea and the Indian Ocean, following the example of Arab dhows. Each navigator knew the time and strength of the monsoons and pushed his vessel before them, as their fathers had done for centuries. Ancient Roman and Indian culture knew of one another and traded with each other because of this constant ocean wind current, far stronger than any sea current.

In pre-historic times oral faith tells of travel via the wind over distances long and short. Genetic and biological evidence points to the discovery of islands across the Pacific, even Australia and Indonesia, via small frail vessels dependent upon the wind. These earliest nameless explorers were perhaps the greatest of all wind voyagers. Either by accident or design, they populated thousands of miles of open sea without knowledge of destination. They populated an island and moved on. Wind was their *sub rosa* proof, their Golden Fleece, their lodestar.

Local knowledge of a riverine passage upstream against the current via the wind or trans-oceanic knowledge acquired over centuries of trial and error, both are a direct result of the wind.

Concave shapes encourage pressure differentials. Wind is pressure differential, across space and time. A steeply walled valley may have strong thermals lasting longer and rising higher. The European griffon we encountered earlier lived and bred and died in the craggy outcroppings of the mountains of Crete, as its committee cohorts do in Spain, Greece, Italy and France. It rose into its food seeking gyres with the morning thermals, as did the paraglider.

The geography of wind is complex. Surface roughness dictates wind speed to a certain degree. Wind runs easily over smooth surfaces, like oceans, lakes, and rivers. It is corrugated over hillocks and craggy terrain. As a breeze approaches a coastline, it accelerates during the day. The land is warmer than the water, drawing in air molecules. Interruptions like forests, cities, and mountains radically alter the flow of air. This is one of the forces maximizing wind shear, or speed.

Prevailing winds (constant winds from a quarter of the compass) run for hours, days, even months. The monsoons of the Indian Ocean are reasonably predictable in intensity and duration. They are fairly, but not entirely, reliable. The summer wind comes down from the Himalayas, warming as it crosses the Indian subcontinent and out to sea, west towards the Arabian deserts. In its warmth, it holds moisture which precipitates as rain. In the winter, as the sea retains its warmth longer than the land, the flow reverses, moving inland, easterly. It is a dry, tepid wind.

The high and low pressure systems—fine weather and foul—are a result of the Earth's rotation. The Coriolis Effect turns a large mass of moist air upon itself, creating a storm. The storm (low pressure) can be a quick summer thunderstorm or a massive hurricane. Storms move across an area quickly, often at 15 - 30 mph. They are sometimes welcome but often a disaster. The pleasant high pressure systems can last for days or weeks, stationary over a large land mass, sucking up warm air from close to the ground and slowly swirling it upwards and out. These are the halcyon days of summer. Little or no breeze for weeks on end. They can be good for well-watered crops and disastrous in drought conditions. They draw the warm, moist air up and out, spinning it at great altitude off to distant lows. The new storms are formed from this descending airflow and the cycles continue, infinitely across nations, across epochs.

Wind speed varies with the height above ground. The shear can be measured very effectively across wide swaths of landscape. It rises with height. The rate of rise lies within a range of height above the ground. Within that range is the best height for a particular wind turbine size. It will become more complex, very quickly, but for now, accept this simplistic description.

Wind speed varies with time, too. Gusts and williwaws come out of nowhere, spilling wind from on high. They get stronger with the approaching storm and the filling afternoon. They billow in crest and trough during a storm. Wind turbines today can change their surface area to maximize, or minimize, wind pressure exposure. Too much pressure will destroy the blades and rotor. Too little will not *spin the dial.*

Wind varies with direction over time, short and long. We saw the biannual monsoons. As a low pressure system approaches in the Northern Hemisphere, the wind fills in from the southwest. As the storm passes overhead, the wind veers from the north and then backs to the northeast after encircling the compass rose. Turbines have to react to these directional changes. Directions can change with seasons as well. The cool fall days encourage a northern drift to the wind direction. High altitude wind systems bring colder air masses from Arctic regions, turning cool to cold and rain to snow. The albino effect of the snow maintains the cold, encouraging more. Only the sweep of the planet around her Sun draws the warmth back to the land, forcing the winter aeolis back to its Arctic home.

Each of these gradients—velocity, direction, tenure, shear, and duration—affects the functioning of a wind turbine. Today's machine is both huge and complex. Their size exposes them to these differentials across their diameter and height. They must be massive. With a single foot etched deep into the earth, their design must be the perfect complement of strength and sleekness. The noise often heard from a wind turbine is the result of some design imbalances. Fatigue failure at the blade root and shaft bearings is a constant danger. Changes in velocity rapidly run down, or up, the electric power output. These voltage surges wreak havoc over a power grid.

You have seen the wind, of course. Wind born particles illuminate the invisible power. The swirl of snow as it rises from your driveway or crests Mt. Everest is the wind. The driving edge of sleet or rain lashing the school bus or your boat is the wind. The particles make it visible. In extremis you have seen the rotating vortex of a tornado or wind spout. You see not the wind but what it carries—the dust and detritus of the fields and shallow seas across which it moves, at a speed that surprises you.

This wind can be measured. Scale brings wind some degree of understanding, if only by defining it. This is a hand drawn illustration of the Great Blizzard of 1888:

15

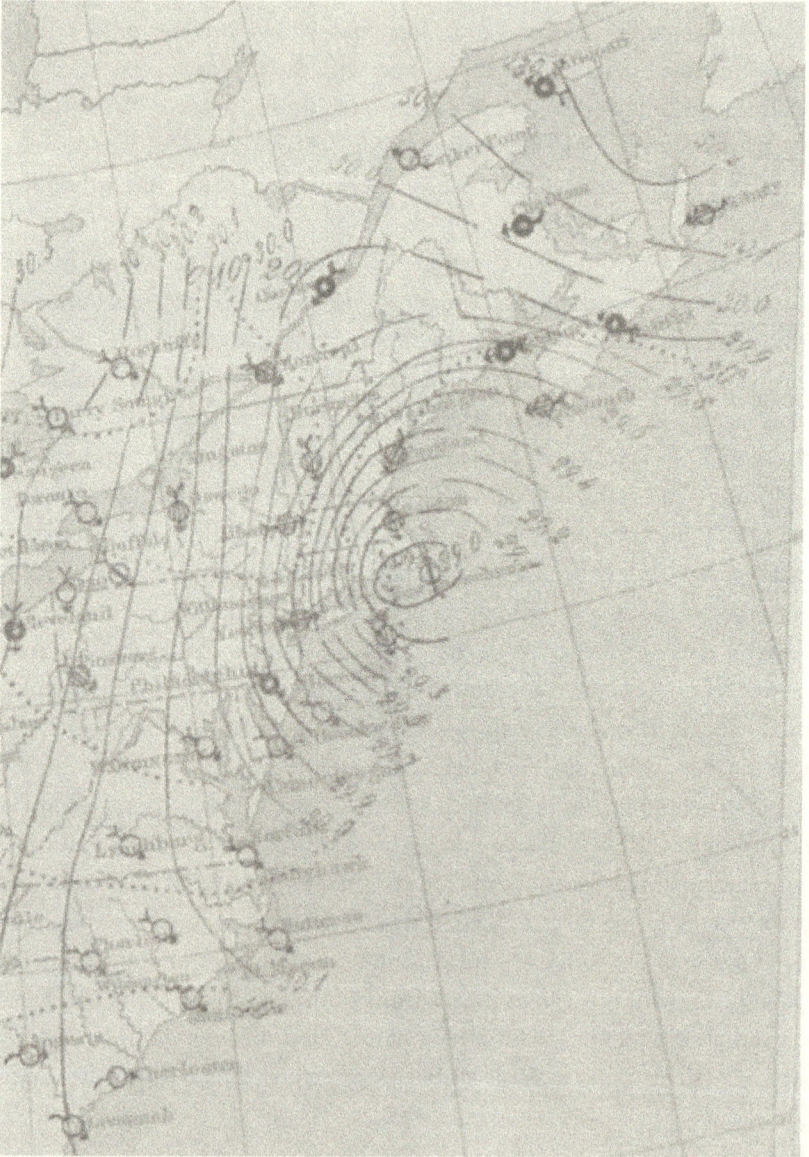

The lines define pressure gradients. These gradients are also the measure of wind speed. The closer together, the harder the wind is blowing and the quicker the wind strength increases. The visual tells the story. The entire Eastern Seaboard of the United States was under the cloak of this super storm of the late 19th century. The devastation of this late spring snow was enormous. 40 inches of snow fell in New York City, with 58 inches upstate. Snow drifts buried three story buildings. 400 fatalities were recorded. It remains the coldest March on record. The storm was also indirectly responsible for the first subways, as the railroads were under snow drifts for eight days. Flooding drowned much of Brooklyn. The stock exchange was closed for two days. All this devastation was the result of the wind.

The gradients in this image are written curves of wind shear, of pressure. This wind pressure is what wind turbines need to work. They have ideals as well as maximums and minimums within which they can function. Before we get to wind devices, let's spend just a few moments trying to get a feel for the extremes of wind and of the weather associated with it.

The Beaufort scale is the most common professional wind scale in use.

In 1810, before accurate wind meters, Admiral Francis Beaufort invented a simple scale to measure how hard the wind is blowing by looking at its effects on trees and water (and umbrellas). Today, shipping forecasts still measure the wind by the Beaufort Scale.

Admiral Beaufort Scales the Gale

art by Rupert Van Wyk

Beaufort Number	Wind Speed (miles per hour)	Official Name
0	0–1 mph	Calm

Smoke rises straight up. Sea is flat and glassy.

| 1 | 1–3 mph | Light air |

Smoke drifts. Sea has scaly ripples.

| 2 | 4–7 mph | Light breeze |

Leaves rustle; wind felt on face.

| 3 | 8–12 mph | Gentle breeze |

Small twigs shake; flags flutter. Sea has small waves with scattered whitecaps.

| 4 | 13–17 mph | Moderate breeze |

Small branches shake; paper blows about.

| 5 | 18–24 mph | Fresh breeze |

Small trees sway. Sea has whitecaps, some spray.

| 6 | 25–30 mph | Strong breeze |

Large branches shake. Difficulty with umbrellas.

| 7 | 31–38 mph | Near gale |

Whole trees sway.

| 8 | 39–46 mph | Gale |

Twigs break; hard to walk against the wind. Sea has high waves with breaking crests, spindrift, and spray.

| 9 | 47–54 mph | Strong gale |

Branches break; roof tiles blow off.

| 10 | 55–63 mph | Storm |

Small trees blow over; chimneys blown down. Sea has high tumbling waves with heavy impact.

| 11 | 64–72 mph | Violent storm |

Serious damage to trees, roofs, and buildings. Foam and spray cover much of the sea. Severely reduced visibility.

| 12 | Above 73 mph | Hurricane |

Widespread destruction. Large trees uprooted, windows break, roofs torn away. Huge waves; sea completely white with foam and spray.

text and art © 2011 by Carus Publishing Company   a s k   19

Many thanks to Rupert Van Wyk for this humorous illustration. For sailors the Beaufort is the standard. We describe our venturing offshore in terms of Force. "I was in a Force 10 for two days 90 miles off the California coast" (a true story). Wind blowing at that speed (the pressure gradient being that significant) is truly unbelievable. You have to shout into your mate's ear to be heard. The sea is ripped, shredded, by the winds. They come from different directions, sometimes building new waves out of nowhere, tumbling your boat sideways, down into a trough, turning on her side until the mast is under water.

Wind turbines need at least a Force 3 to begin to spin. Above Force 9 they are flared to spill the excess wind. Between these two capacity extremes, turbines work their magic. The energy source is free and unlimited. If it blows for days at 40 to 55 mph (actually kph, knots per hour), all the better for wind farms. This is their element. As long as they don't drop a blade, they will perform at peak capacity, at load capacity.

We will examine wind farm placement later as we review external impetus into siting patterns across several nations. European countries share a common wind heritage from the North Atlantic. They share common wind speeds across the western part of each continent. Surprisingly, wind velocity differentials between Scotland, Wales, Holland, Denmark, France and Germany are very similar. A few hours' time lag between sites is expected as a weather system approaches Europe. As the system passes over the continent, the various wind farms report similar lengths of time during which similar wind speeds are enjoyed. The opposite is also true. When a high pressure system settles down over Europe, all wind stations are idle, often for days or weeks. Of course, these highs often occur during peak demand: cold winters and hot summers.

The ideal weather conditions for wind turbines happen to lie across the midsection of America—tornado alley—and across the western mass of the Eurasian continent. Steady, strong winds blow, the ideal fuel for this 21st Century technology of wind farming.

At the extreme the photo below is a Force 12 image from a VLCC—a very large crude carrier, 800 feet in length. Her deck measures 80 feet across. The wall of water just passing beneath her bow is estimated at 60 feet in height. That is the height of a five story building. The vessel climbs the wrack ponderously. You don't see the next frame, where she falls off the top as it runs forward beneath her hull. This is the most dangerous moment for an ocean going ship.

BEAUFORT FORCE 12
WIND SPEED: 64 KNOTS

SEA: SEA COMPLETELY WHITE WITH DRIVING SPRAY,
VISIBILITY VERY SERIOUSLY AFFECTED. THE
AIR IS FILLED WITH FOAM AND SPRAY

Beaufort scale image from http://www.crh.noaa.gov/mkx/marinefcst.php

As we can see, measuring the wind may give it understanding. It does not give it gravitas. You just have to be there to feel it...

Let's do some wind turbine math. How do we get energy from wind? How do we translate pressure to electricity, which is simply another form of pressure?

The 'E' energy potential (the kinetic energy) from wind is a function of the 'm' mass of air crossing a 'T' rotor blade.

$$E = m(T)$$

Doubling the circular area of the rotor blade (T), its diameter doubles the mass of air (m) crossing the circle scribed by the blades. This doubles the potential energy (E) made available. Hence, the gigantism of the newest wind turbines. Size matters here (up to a physical point).

The next equation is less obvious. The 'E' energy carried by a mass of air is proportional to the 'f' flow rate cubed.

$$E \; \alpha \; f^3$$

We won't go into the mathematics. Ask your 7th grade science student for a refresher course in basic algebra; it was fun for them to learn and they may enjoy teaching you today! We shall simply describe what this means. If the flow rate (speed) doubles, the potential energy carried increases eight fold (2x2x2 = 8). Conversely, if the speed drops by half, the amount of potential energy available drops to ⅛ (1/2÷2÷2 = ⅛).

A third formula is known as Betz Law.[16]

$$P_{\text{wind}} = \tfrac{1}{2} \bullet \rho \bullet S \bullet v_1^3$$

We do not wish to open this cryptogram. The footnote will drive you to the translated writings on fluid dynamics by Dr. Betz. For this discussion we want to know this simple fact: a maximum of 59.3% of the kinetic energy available from a wind source can be extracted by a wind turbine. Mechanical interference adds to energy losses via drive train, alternators, etc. They reduce this further to a maximum of about 30%. This happens to be quite similar to coal and gas-fired power generation systems, which run between 35% to 45% efficiency. Efficiency is a trite word. Effective makes more sense. The effective translation of kinetic energy from wind through the turbine is about 30%.[17]

The actual amount of distributed power is a result of the blade spin and how efficiently they turn. At ideal wind speed this can be as high as 40%. It is often zero. The wind blows intermittently. Its speed changes. The result? 6% to 8% of available energy is translated down the wires to distribution channels—high tension wires strung across the prairies and mountains.

Our final formula is probably one of the first that you learned.

The area of a circle equals 3.1416… times the square of the radius. Doubling of the area scribed by the wind blade circle squares the kinetic energy available from which to draw. The formula simply tells us that a small increase in the rotor diameter (twice the radius) increases potential power significantly. The rotor diameter tells you more about the turbine size than the generator itself.

For example, a rotor diameter of 11 meters, or a *swept area* of 1,000 square meters, has a nameplate rating of 300-400 kW. Double the swept area to 2,000 square meters and the power increases to 500-700 kW. Double it again to 4,000 square meters and the power increases to 1,000-2,000 kW. Thus, the large sweeps of blades in excess of 150 feet in length. This sweep allows for far greater power potential. At these largest sweep sizes, the turbine puts out the equivalent power of a diesel locomotive.

A final clarification is in order in the science of wind. What is the difference between energy and power? Energy is the capacity to do work. Power is the rate at which work is done. Watts, kilowatts, and megawatts are measures of power. Kilowatts per hour or megawatts per hour is the rate at which the energy is delivered; it is the work delivered from the power source. This is what you pay for with your electricity bill each month.

Here we turn the page. From the science of fluid dynamics, let's go to the engineering of wind turbines, yesterday, today and tomorrow. How are turbines designed and produced today? What are the power distribution issues facing wind farms and their tie-ins now? How do four nations define and face their critical energy needs of tomorrow?

CHAPTER TWO
OSPREY

High winds blow on high hills. — Thomas Fuller, *Gnomologia*

Plunging into the water feet first, its masked eyes tracking its underwater prey, the osprey is also known as a fish hawk. They are among the largest of Buteo raptors stretching two feet in length and having a wing span in excess of five and a half feet. With scaled feet, oiled wings, and reversible toes that enable a sure grip, the osprey is a deadly fisherman.

https://www.youtube.com/watch?v=BN3odzZILdc

The osprey is as universal a bird as planet Earth has divulged. Its migratory patterns are enormous, taking it full swing through both northern and southern hemispheres over the course of a year.

Technology

Let's examine wind energy technology, production, distribution and use. The industry has grown enormously across Europe and America since the 1980s. What has driven this growth?

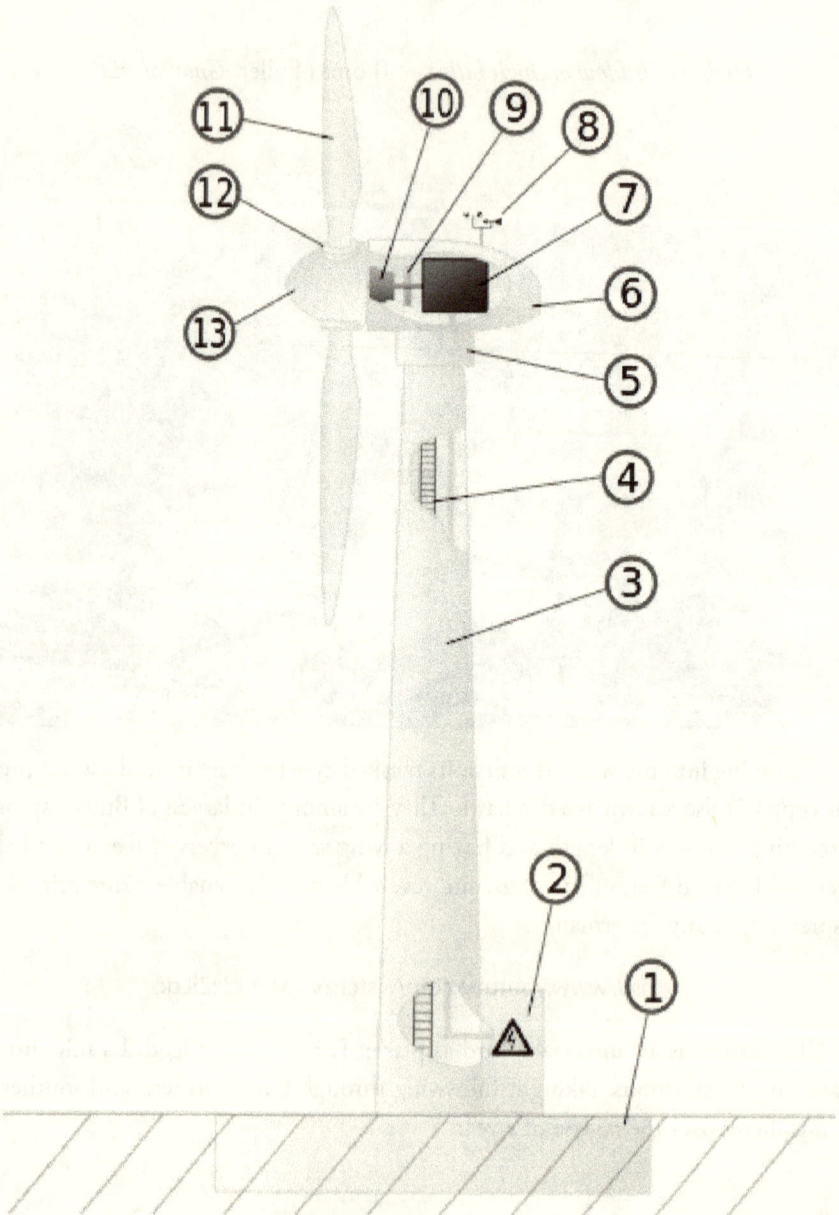

Illustration of a wind turbine

First, let's take a very brief look at the workings of a wind turbine and its tower. These have changed significantly over the previous thirty years. The Danish have led the design field.

1. Foundation
2. Connection to electric grid
3. Tower
4. Access ladder
5. Wind orientation control
6. Nacelle
7. Generator
8. Anemometer
9. Brake
10. Gearbox
11. Rotor blade
12. Blade pitch control
13. Rotor hub

How does a turbine work?

Wind turbines first must convert longitudinal kinetic energy into rotary motion, into mechanical energy. Generators convert this mechanical energy to electrical energy—into power. Originally, wind brushing against a windmill pushed the sail blades out of the way as it moved through the diameter of the scribed circle. These blades rotated and the gearing at their base was transmuted through 90 degrees into power to turn a millstone. Today's blades are aeronautically designed to create lift and drag, just as an airplane propeller does. This is far more efficient at the conversion process. Betz would be proud.

Blades are constructed of fiberglass matt and balsa core, epoxy resin, adhesive and steel root, connecting bolts, lightning rods, etc. A twist is designed into the shape for aerodynamic efficiency. The blades are thin and long to enhance performance during low wind speeds.

The blade pitch is controlled so as to maximize energy transfer from wind to rotor. It also dumps excess wind when the velocity exceeds safety margins. Turbines begin to work at about 7-9 mph and have to be shut down above 56 mph. Shutting down means dumping the excess wind by slanting the blades to avoid all kinetic energy transfer. They are set parallel to the wind, like window slates.

The blade *root* is encased in the rotor hub and coupled to a *gearbox*. This increases the slow blade rotation of 10 to 30 rpm, to the speed required for AC power generation, or 3,000 rpm. Either planetary gears or spur-helical gears are used. The former is more compact, while the latter is less prone to overheating and nuisance noise. The *brake* is released once the local wind speed becomes acceptable, about 9 mph, and engaged when these velocities exceed about 56 mph. The anemometer is the wind speed indicator at the rotor hub.

The *generator*, or alternator, is either asynchronous or synchronous. It may have a direct or indirect connection to the electrical grid. Indirect connections require computer control for voltage and frequency to match the grid network's power features. Asynchronous generators do so automatically and are called induction generators. The power generated by the turbine is transformed up to the grid voltage by the transformer.

The *nacelle* is the protective enclosure for the drive train and generator. This may have a helo landing pad on the largest turbines. The nacelle provides the structure and mounting (load transfer) for the large bearings on the wind shaft itself. These may have to contend with massive stresses on large turbines, bearing the weight of thirty tons and more.

The wind orientation control (*yaw drive*) does just that; it keeps the turbine blades facing the wind. An electromechanical drive keeps the turbine in the wind automatically. Wind direction varies in small and large degree. Each puff can come from a slightly different angle. Just ask a sailor as he constantly adjusts his course to the best wind angle as that angle changes. He luffs up or falls off. Direction can also change through the entire compass rose. Recall our discussion on storms. As one approaches, the wind fills from the southwest. As it passes over, the wind veers to the north and then backs to the northeast. The turbine direction must change with these changes in wind direction. If this is not programmed correctly, asymmetrical loads on the blades can significantly reduce power production or destroy the turbine.

The electric grid is connected to each turbine through the power lines to the grid.[18] The power lines running down the tower column have to float so as to never tangle as the nacelle and blade assembly rotate with the wind.

The tower carries the weight of the turbine, lifts it above the laminar ground air flow into the ideal wind stream and resists tilt stressing from the wind on this often 30 story building. It is centered upon the base foundation. The foundation can be either a spread-footing or deep foundation design, depending upon soil conditions, load factors, and turbine weight. The deep foundation contains the base stability plate and several feet of concrete with reinforced steel bars. Tons of cement and stone make up this massive footing. Boltings hold the tower structure to the base structure in a bolt cage. In offshore wind turbines, the footing may be encased within a caisson or sunk beneath the ocean floor.

Siting the towers is the most critical issue because it is the most cost-effective economic consideration. Windy sites with tall towers are the ideal towards which operators strive. Initial costs must be expensed over a 15 to 30-year lifetime of usage and actual energy output. Actual life spans of these turbines is considerably less than the amortization.

A Short History of Wind Technology

Wind has been used longer than the written word. Images convey its use well before Egyptian or Chinese calligraphers put these images into form and forms became words. The discovery and exploration of much of Southeast Asia and all of the South Pacific had to have been accomplished, by accident or design, with wind power.

Heron of Alexandria recorded the first known wind use for power.[19] By the height of Persian power in the 9th century, windmills were in use across the Middle East and imported to India and China. The first writings on European windmills begin in the 12th century, but their use certainly predates these records. Grinding of grain and draining of swamps were early uses.

Modern times for wind energy began with James Blyth who converted sail driven mills to electricity in 1887. A decade later Poul la Cour, a Dane, proved the validity of today's turbine design. His Society of Wind Electricians was founded in 1904.[20]

Wikipedia, open source

By 1956, a student of la Cour, Johannes Juul, built a 200 kW ensemble, further dictating— and limiting—the three-blade design for wind turbine power generation. The U.S. Department of Energy, via NASA in 1975, started the first government initiative to fully implement a wind-centered power generation facility.

Catastrophic is the best description for many of the wind turbine experiments in the 1970s and 1980s. Spurred on by massive tax credits, the interest of the new environmental movement, and the drive of politicians for new babies to kiss, the new wind beasts of the time often followed their rocketry cousins in design disaster.

Initially, smaller was better, as the feds tried without success to promote small turbine design at Rocky Flats in the 1970s. The manager of the facility sums it up quite nicely: "We tended to be blinded because windmills had been used for more than 1,000 years. We thought the technology was there and all we had to do was bring it into the 20th century."[21]

An early failure in Sandusky, Ohio of the Boeing MOD 2 unit led to the *New York Times* headline that read, "$1M for only 30 Hours of Work".[22] In 1979, the DOE tried to use wind power to reduce water use along the upper dams of the Colorado River with a planned 50 turbines offering 100 MW capacity. Congress approved two units: a Boeing MOD2 unit of 2.5 MW and a Hamilton Standard of 4 MW. By 1982 they were installed. Catastrophic failure followed after sporadic power generation over 18 months. They were sold for scrap metal in 1987 for $13,000 and $20,000. O&M killed the project.

These early experiments often exploded on the launch pad, as the Alcoa 500 kW did after two hours of running time in April of 1981, just as the California Energy Commission was opening its energy conference in Palm Springs. Bendix failed with a 3 MW creature trimmed down to 1.2 MW. United Technologies failed with their prototypes in Wisconsin the previous year. In June, the Columbia River Boeing unit wrenched its innards and tore apart. In 1983, federal testing of 29 out of 32 units suffered from major mechanical failure.[23]

The industry is quick to condemn the federal agencies for these failures. Rigid solutions were rarely changed. Alternatives were mocked. 20 MOD2s were produced, and none succeeded at the task. At the time young men were pursuing their dreams of building the energy source for the future. The hydrocarbon energy world had been shaken twice by Arab boycotts. Paul Erlich was ruing the future destruction of the industrial world at its own dirty hands. The world was going to end under a sheet of a new Ice Age, as carbon dust settled over the lands, reducing heat and light, and polluting our eyes and lungs. We were consuming our own flesh in the form of the destruction of Earth's few remaining vital resources.

The Europeans took a different tack. They built strongly engineered, craftsmen designed units. Without governmental support, the Danes had built smaller, functioning wind turbines since the war. Vestas is the survivor today, having captured a significant share of the current turbine market with reliable machinery. As late as 1996, a wind author was still uncertain that the Danish model would ultimately win over U.S. engineering skills.[24] Time has proven otherwise. Vestas gained market share during the recession and the new Administration's support for alternative energy sources. Their design became ubiquitous, the model for all competitors. Recently, they have struggled as GE and Siemens have taken market share.

While the monopole design reigns supreme, today's manufacturers have changed leads in the race for wind. Like massive racing yachts powered with infinite money, these new energy companies plough tax credits and guaranteed pricing into bigger and bigger designs, all on the same theme, monopole, three blades. Bigger is better, taller is more powerful. It is all about size.

In England, the story is similar. They are a few years and a few billions ahead of us. They do have a concentration of wind in the North that is not as relatively distant as our Midwest is for the coasts. Here is the comparative story from London, just this year:

> When Professor David MacKay stepped down as chief scientific adviser to the Department of Energy and Climate Change (DECC) last year, he produced a report comparing the environmental impact of a fracking site to that of wind farms. Over 25 years, he calculated, a single "shale gas pad" covering five acres, with a drilling rig 85ft high (only needed for less than a year), would produce as much energy as 87 giant wind turbines, covering 5.6 square miles and visible up to 20 miles away. Yet, to the greenies, the first of these, capable of producing energy whenever needed, without a penny of subsidy, is anathema; while the second, producing electricity very unreliably in return for millions of pounds in subsidies, fills them with rapture.[25]

Current Installed Wind Capacity (MW)

Washington 2,808
Oregon 3,153
Idaho 973
Montana 645
Wyoming 1,410
Nevada 152
Utah 325
California 5,587
Arizona 238
New Mexico 778
N. Dakota 1,681
S. Dakota 763
Nebraska 459
Colorado 2,324
Kansas 2,713
Oklahoma 3,134
Texas 12,214
Minn. 2,987
Iowa 5,133
Missouri 459
Wisc. 648
Illinois 3,568
Ind. 1,543
Mich. 988
Ohio 428
W. Va. 583
Tenn. 29
Penn. 1,340
New York 1,638
Vermont 119
Maine 431
New Hampshire 171
Mass. 103
Rhode Island 9
New Jersey 9
Delaware 2
Maryland 120

Data is from the American Wind Energy Association Third Quarter 2013 Market Report.
http://www.awea.org

Total: 60,078 MW
(As of 09/30/2013)

Alaska 61
Hawaii 206

Wind Power Capacity
Megawatts (MW)
1,000 - 11,000
100 - 1,000
20 - 100
1 - 20

U.S. Department of Energy

:NREL
NATIONAL RENEWABLE ENERGY LABORATORY
20-NOV-2013 1.23

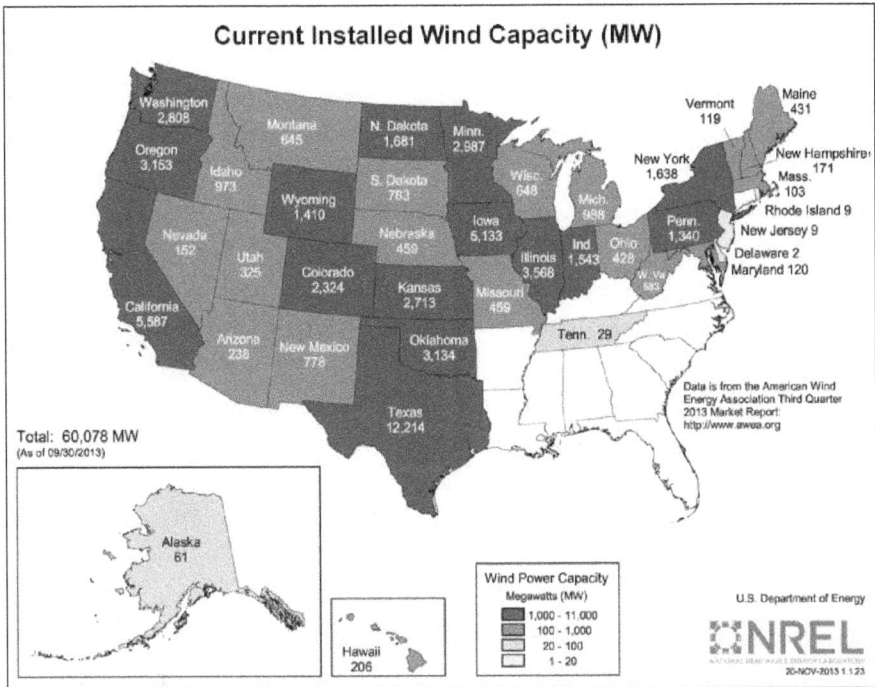

Federal subsidies have been a constant, addictive irritant for the industry. The industry is constantly making pearls out of the federal silk purse. The purse—federal grants, subsidies, tax credits and regulatory fiat—has only opened wider with the passage of time. Failure has bred success in funding millionaires and billionaires in smoke-filled rooms with their pipe dreams. The addiction has spread and all thought of the economies of wind energy production have dissipated.

Rent seeking prevails. This is the securing of guaranteed capital from government sources via legislative and regulatory fiat. The diktat of the environmental proletariat demands capital conscription in ever larger amounts. While wealthy hydrocarbon barons are excoriated for their bilious billions, other billionaires are heroes for leaping from the hydrocarbon frying pan into the wind fire. The difference? They get paid for the pyrotechnics by a grateful government.

In the 1980s, these new tax resources were the glimmer in the newborn father's eyes. Invest in wind and get 100% of your capital back from the IRS in one year! The tax credit concept was developed under the Brown administration in California in the early 1980s. Wealthy investors would receive capital back from the tax collector for their investment in alternative energy solutions such as wind and solar. Federal legislation encouraged the sale of electricity from small cogeneration and alternative energy producers and the concomitant purchase of their generated power at preferential rates.[26] The monopoly on power production enjoyed by utility companies for nearly half a century was broken. The small producer was to have the advantage over the large. Environmental logic was to hold sway over economic hegemony. Good premise, erroneous consequences.

California discovered, much to its surprise, that it had more wind energy potential than it realized. Mountain passes begat the new gold rush of the 1980s. Wind was gold. A 10% target for wind energy sourcing for California's future projected power needs was feasible. Lifting water when wind was strong and storing the kinetic energy in the water, to be cascaded later through hydroelectric facilities was the new dream team combo. Only 29,000 4 MW wind turbine were needed. Best of all, aesthetics, land use and air pollution issues were negligible. $5 billion was the goal. Tax credits were the players. Cheap financing greased the action. Municipal bonds would support the infrastructure demands.

The rush was on. Tax credit partnerships only needed a pen and a few pages of memorandum to get a check. Commissions flowed like water to the investment advisors and their brokers. Broker, in this case, is the operative word. None of these schemes paid off in cash. The IRS investigated their abusive structures in the mid-1980s and fined a few wholesalers. Altamont Pass to the east of San Francisco, Tehachapi to the north of Los Angeles and San Gorgonio Pass to the east were the ideal windy sites.

Ranchers gained new revenue from leasing their cattle spreads to the turbine cowboys. European turbine manufacturers jumped on the wind bandwidth. Local politicians made good on their promise to cut the ribbons on their new free energy sources.

http://www.youtube.com/watch?v=aU9MHNL9AQk

By 1985, 12,553 turbines were spinning for dollars in the deserts of sunny California, all supported by wealthy investors and their new chums, the IRS. Installed capacity was 911 MW—96% of the U.S. wind capacity. The end of 1985 also saw the demise of the new tax credits. The industry took a flyer, manufacturers took a tailspin, and investors took a bath. The IRS took a closer look.

The lobbyist for the wind crowd, AWEA, took a breather from sunny Palm Springs for more than a decade in protest to its suddenly chilly views toward wind energy. Punishment was swift to the unbelievers.

While the industry blamed the mayor, Sonny Bono, his constituency was more at fault. They quickly turned their backs on the new credits, jobs, and taxes generated by the white towered beasties along I-10 out of Los Angeles. Many towers were dysfunctional. Some blew up. More than a few broke down or never ran in the first place.

http://www.youtube.com/watch?v=-YJuFvjtM0s

The sands of the desert destroyed the mechanical guts of these air creatures. Small rocks were found embedded in a turbine 100 feet in the air, freezing the joints forever. 12% production capacity was the best these creatures could offer on the best of days.

Several designs were tested, installed and failed. Law suits followed. Companies went out of business. Utilities tore up contracts. But the best was yet to come.

The Tejon Pass was the next target for turbine technology. Zond Systems had built out the Tehachapi wind facility with 1,338 turbines of 120MW installed capacity. They added 342 Vestas 225kW units in 1991, controlling 15% of California's wind capacity. Their next target was the Tejon Pass between Los Angeles and the Central Valley. 4,458 turbines were to be installed. A fund was set up to share the wealth with local residents. A consortium of environmental groups was enlisted for support: Sierra Club, Audubon Society, World Watch, the Rocky Mountain Institute, Zero Population Growth, and Ralph Nader. Who could beat such a mighty team?

The mighty condor. It could and it did. Our avian friend, so near to extinction, had its last remaining habitat right inside the turbine wind farm. Promises to monitor flight schedules for the birds and turn down the blades were scoffed at by the local Audubon Society's Linda Blum, "Monitoring tended to be fairly unreliable."[27] The greatest raptor in Earth's skies was joined, through threats of extinction, by its cousins, the eagles, peregrine falcons, and Swainson's hawks. The planning commission rejected Zond's proposal unanimously.

We should recall this lesson today. These birds remain at risk. The risk is far greater. The results are far more egregious. The assault is far more ingenious. The threat is from the federal government, once the protector of wildlife. Recall the words behind the moniker—the U.S. Fish and Wildlife Service. Their mission is *conserving, protecting, and enhancing.*

We shall see that this is yet another broken promise, yet another federal, regulatory, and industrial lie. The swamps of DC are infested with vermin that maim, kill, and destroy wildlife.

You must observe that in the case of wind turbine design, government backing has defined the limits of creativity. After more than 40 years and $80B in subsidies, the industry has consistently failed to stand on its own two capital feet. It continues to depend upon tax subsidy for existence. Irrespective of its stated energy efficacy merits, these economic marshlands may never be drained of rent seeking pseudo-industrialists.

Of course, a few problems attend every new technology. View them here:

http://youtube.com/watch?v=YCvf61_ebgM&list=PL3A07E3FD993AD6B8

The design drives the manufacturer of today's wind turbine. Let's examine the production of wind in more detail.

CHAPTER THREE
EAGLES

The film below is designed for youth to learn of other cultures, other ways of life. Yet it also shows important cultural values for you. The images of the Central Asian frontier, the vast desert and mountains of Tien Shen, the people who love this land—all of these are critical to a better understanding of humanity and our interaction with avian cultures. Sharing this with a youth in your family is an immense treasure:

https://vimeo.com/ondemand/eaglehuntersson/48561944.

For a look through the eyes of a brilliant black and white photographer, enjoy these few minutes with John Delaney in Mongolia finding and following the hunters, man and eagle:

https://vimeo.com/3583993.

Many locales are losing their eagle populations, or have lost them entirely. Britain, Scotland the Slovak Republic and the Czech Republic are reintroducing goldens into the wilds:

https://vimeo.com/138591087

https://vimeo.com/138591089

https://vimeo.com/22117747

A bird's eye view from a young Golden Eagle on the fringe of the Taklimakan Desert in western Mongolia, an area visited by the author:

https://vimeo.com/75234192.

The most unusual story of golden eagles and humans goes by the name Ashol-Pan. She is a Mongolian teenage eagle huntress. Enjoy her story from the BBC:

http://www.bbc.com/news/magazine-26969150.

For the hunter in you, enjoy this piece from the BBC. The action is fast, the scenes exciting and graphic. A young golden, with GoCam attached, takes a fox. Her traditional reward? The lungs. The hunter keeps the pelt for his family to make hats and gloves for him and his siblings.

http://www.bbc.co.uk/nature/humanplanetexplorer/environments/ mountains#p00dwd5x

© 2014 BBC. Reproduced under licence from BBC News (bbc.co.uk).

In these cultures, the golden eagle is a close family member. She has a simple role: kill foxes and wolves. Wolves attack the horsemen's herds. Eagles hunt and kill wolves. With a wingspan in excess of seven feet, these birds will dive down upon a lone wolf from behind with talons stretched forward. The attack often breaks the neck. If the wolf survives the first onslaught, it still has to fight in close quarters with a skywalker much larger, not in weight but in extent. She mantles her wings around the victim completely covering it while holding the snout shut in a death grip with four-inch-long razor-sharp talons designed to crush bone. Dazed and disoriented the wolf has but a few moments to try to escape or bite. The beak comes down and rips open the throat. All is done.

Her keeper rides like the wind to the scene of carnage. He carefully puts a fresh piece of raw meat in her face to lure her away from the slaughter. The death is important to all members of the society. One less wolf to contend with during the long winter. A well trained killer advances her pride within the human tribe. The wolf pelt will clothe the family. The carcass will feed the birds and dogs.

I held her on my arm. She showed no animosity towards me. Yet, I knew she could as easily pluck an eye before I could blink. Savage beauty on the steppes of Central Asia. This is a world of survival. The landscape's beauty is deep, iridescent. The people are warm and gregarious. Two boys had ridden up to my truck an hour before, bareback on a small roan. They asked me in Russian if I'd like to see eagles, their eyes laughing at my English response. They were entertained and were the entertainment for us. Roughhouse boys with tousled hair and bright eyes.

http://ibc.lynxeds.com/op_video/57001/embed

Production

Construction of the massive monopole towers is an engineering wonder. Watching the form take shape over the course of a day in a quiet dell in England leaves you to wonder how they figured it all out. The footing is dug and built first. Assembled like a children's game of blocks, the sections of the tower are lifted and piled atop one another. The nacelle and generator complex are added. Finally, the blade components are coupled together and lifted as one piece.

http://www.youtube.com/watch?v=r0DZUDQyw_0

http://www.youtube.com/watch?v=iJ89Aw3Y86Q

The largest wind turbines today have a rating of 7.5 MW (megawatts). This is the *nameplate capacity*, or *installed capacity*. It is the full load capacity of an energy-generating facility running 24/7/365. Under ideal conditions, installed capacity is "what this baby can do." The *load factor* is "what she delivers." It is the actual amount of electricity generated over a time period, compared to what it would have generated under ideal conditions running continuously at maximum capacity. It is also known as *capacity factor*. In the U.K. it has averaged 26% of nameplate capacity. The best figures come from New Zealand, with as much as a 40% capacity factor. In the U.S., figures of 28% to 36% are frequently used without substantiation. Many writers have used a U.S. figure of 20%.

While this apparent inefficiency is huge, it is considered in financial and regulatory projections. It is rarely referred to by the media, which promote the larger figure for reasons of their own. Communication is relative today. What is said is relative to what the speaker wants you to hear or know.

Capacity credit is the next phrase of importance. It represents the statistical ability of wind to replace conventional power sources—usually coal, natural gas, or nuclear. It is the amount of conventional power generation that could be shut down without severe load impact to the grid.[28] The wind energy industry makes this case, as the footnote shows. It then conveniently ignores the low capacity credit figure by stating that it is irrelevant. The true value is in fuel and carbon emissions reductions.[29] The purpose of the set piece is to reduce, replace, or eliminate carbon dioxide in the atmosphere. Confused? Of course you are. That is the purpose of their exercise. Confound with words and figures.

By early 2013 the global nameplate capacity was 282,482 MW, or 282 GW (gigawatts) of wind energy. Capacity credit in Europe today is estimated, by the CEO of E.ON (the largest European wind farm operator) at 6% of installed capacity. Let's do the math:

6% of 282,482 is 16,948 MW
Convert to gigawatts: almost 17 GW

Thus, the entire global wind energy production might reliably replace 17 GW of conventional energy production. The EIA estimates[30] global energy generation for 2010 of 20.2T KWh, or 2,020 GW. Of this, 1,600 GW are from conventional sources. Wind can replace approximately 1.1% of energy generated globally in terms of capacity credit. Reality hurts.

Losses from wind turbine power production are legion, even when the turbines are functioning as designed. These loss assessments are an important aspect of design, site and construction profiling. Here is a table of potential mechanical losses:

O&M	2-5%
Electrical	2-4%
Turbine performance	2-5%
Environmental	1-3% (weather, wildlife, lightning)
Curtailment	1-3%
Tremors	1-3%
Wake (flow interference)	5-15%

Operators build a 10% *loss profile* into their financial projections, much as a real estate management firm might assume a 10% vacancy rate. These loss profiles become bankable and create an audit trail for potential investors. They are, of course, pure guesswork—*quantifiable uncertainties*. One example should suffice. Wind energy is stated as a function of the quadratic of wind speed, on paper. In practice it is the cube of wind speed. This is a huge mathematical difference that expands with estimates. Insignificant alterations in real wind speed at a site vs. observation-based *estimated wind speed* over several years can make or destroy all other figures in the bankable resource estimate. Good luck with the guesswork.

A supply of electricity with a high capacity figure is *despatchable*. Conventional power generation (coal, natural gas, hydroelectric, and nuclear) is eminently despatchable. It is available at any time, in a very short period of time, to satisfy demand by either increasing or decreasing production.

Production from a low capacity source is less despatchable. Clearly, wind is neither readily rampable nor reducible. It is challenging to turn it on and off at will. The power source is at the will of the gods. Gearing technology attempts to regulate this variability at the lower end of the wind speed function. 7 mph through 12 mph speed differences can be mollified with gearing to affect greater regularity. A greater variability in wind speed, or a speed range spread across higher wind speeds, is currently unmanageable. It can be turned off, but not regulated.

Even this is less important than the *unpredictability* of the power source. Recall that wind provides kinetic energy. With the vicissitudes of *wind direction, velocity, and constancy*, it is utterly unpredictable. You cannot reliably or accurately predict the power generation from a turbine or wind farm with any degree of precision. These three elements of wind power sourcing change by the second, minute, hour, and season of the wind. Did we mention it is diurnal as well? Its strength and direction depend on the time of day.

Backup power generation, *spare capacity*, has to be available at all times. Let's look at the basics of power distribution. Grid physics requires the functional flow of electricity of no less than 85% of *peak demand*, while regulations typically require 110% or more.[31]

Another calculation for capacity is LOLP—*loss of load probability*. It is done twice—with and without wind capacity—then conventional plant capacity is debited. The result of such a computation is a wide range of wind energy efficiencies: from 2% through 40% or even higher. That is, systemic inefficiencies further reduce generational efficiencies. Resulting power generation is even less than capacity crediting figures, which are far lower than the load factor which, in turn, is lower than the installed capacity. Bayem's ideal 59% has degenerated down to 10 to 12% potential.

The environmental value of wind power generation is also important. This is the coat rack upon which all these mufflers and great coats are hung. The idea is to prevent the destruction of the planet by an excess of CO_2. All else pales in comparison.

Calculating this value is another form of guesswork, just as LOLP. Ostensibly, it is the value of reduced emissions of SO_2, NO_2, and CO_2. This reduction results from the perfect substantiation of wind (and solar) for hydrocarbon-sourced power generation. It is also the value of the social benefit of differing energy production sources. Such a calculation depends upon a wide number of variables, a wider choice of application of these, and the *parameterization* (choice of limiting factors in an equation) of these educated guesses. Below is a partial list of queries that go into beginning to answer this equation. The equation has become a polynomic multi-variable series of known and unknown variances across a spectrum of N dimensional space.

In layman's terms? Anyone's guess.

- What is the conventional energy source?
- What are the current emissions from this source?
- What, if any, emissions controls are in place on the system?
- What externalities are considered?

- What is the current cost of the conventional power source?

- How are these costs compared to the wind solution?

- How are they monetized (taxes, regulatory support, etc.)?

- What are the sunk capital and maintenance capital requirements for each system?

- How are these calculated?

- What is the source of capital for insuring these risks?

- If governmental, how is it monetized?

- If insured, what are the risk parameters and how are the expensed?

- What is the monetized value of a clean environment?

- How is clean defined?

- How is social benefit defined?

- Where do wages, capital use, alternative patterns of spending come in?

- Is there a *time value of money* calculation applied?

- What is the whole life cycle cost of each power source?

- Are these calculations fairly equitable across the comparison fields?

As you can see, there is no single way to answer these questions. Thus, it is improbable that any answers concluded are of more than relative value. Relative merits may be debated, but absolute statements regarding their truthfulness are impossible. They become arguments—debating points over a beer on Friday night. Yet the assumed beery answers thus arrived at are thence stated as fact. Regulations are based upon their supposed facts. Quantitative references to these facts are conveniently ignored. *Just trust us on this. The planet will die unless we save it NOW.*

Potential environmental and/or social benefits resulting from wind power generation are absurd conclusions to an impossibly complex equation. Telling the public that wind power today has a cost of only $.05/kWh is absurd. It is, at best, an educated guess that may apply a few minutes each day. At worst, it is simply untrue, yet another broken promise.

Notice that we are not even discussing the present value calculations of future knowledge applied to a task today defined as catastrophic. We cannot know what we cannot imagine. Let's explain with an example.

Instantaneous communication in 18ᵗʰ Europe was impossible. Any problem, such as war, had to endure the *test of time travel*—the speed at which communication moved at the time. That speed can impact decisions made in real, slow, time. The easy example is the American Revolutionary War. Once those shots were fired at Lexington and Concord, the dogs of war were slowly unleashed and no commander could halt their predation. Six years later, a surrender went unnoticed for weeks until London was finally informed that it had lost a continent to a small band of simpletons, rebels with a cause.

The instantaneous communications available to us today are one means of reducing the risk of such an event. While other concerns may reignite the issue, direct and instant communication is no longer of much concern to heads of state. In fact, it can instigate a further problem: how do we keep quiet about our discussions when the NSA, CNN, and Mr. Snowden are listening?

Now, assume your grandson arrives at a solution to the worldwide use of hydrogen as a power source and his sister discovers a way to store massive amounts of electricity for long periods of time at very low cost. In a few years, most conventional and alternative power generation technologies become obsolete. Today's concerns about climate change—and the concomitant aggrandizement of alternative energy sources as a solution to the problem—become meaningless. In fact, they become a complete waste of capital: money, labor, material, and environmental capital.

You can also, alternatively, imagine your grandchildren arriving at these same conclusions, only to be swept away in a tide of cannibalistic war of survival among the few remaining humans on a heated, dying planet, but we'll leave you to your own fantasies.

We can only discuss with any degree of certitude a few of the listed questions. The market values of energy generation and power production, while complex, can be discovered. To do so, we must identify capital costs, O&M costs, revenue sources and types, regulatory mandates and controls, all while taking into consideration current and future breakup values. Conventional power generation firms have done so for decades. They know what their costs are to own, operate, and recycle their plants.

Wind is not free, as many in the industry argue. You must design, build, acquire, install and buy the equipment to harvest wind's kinetic energy. You must convert that into electric power, then distribute that energy into and through the grid. You must try to do so at a profit, even if only a regulated profit. You must maintain and operate your equipment for 20 years or more. You must eventually dispose of the broken bits remaining, including the rare earth metals that killed so many Chinese workers decades ago when they mined and built your magnets. This is life cycle accounting. While the industry demands that oil and gas and nuclear power generation firms perform these calculations, they are surprisingly absent from the wind industry itself. Perhaps this is because it is such a young industry, one that has yet to complete an actual full cycle from creation to destruction…or perhaps there are other reasons. One would think that 40 years was sufficient time to observe the wind industry's maturation. Think again.

Wind power generation systems were initially intended as a supplement to, rather than an ultimate replacement for, conventional power sourcing. How does the constant necessity of backup power qualify as spare? One would think the wind-sourced energy was spare capacity given the many variables just observed. Certainly there is a logic to its use as supplemental power generation, ignoring the avian and human health issues for now. Yet this is not the case.

A few intractable problems immediately surface.

First, you cannot simply flip a switch to turn conventional power generation on and off. It can take hours to start up a coal power plant generating electricity. At best, today's energy efficient, CO2 reducing, natural gas combined-cycle power plant will take half an hour to come on line.

Second, if you try to modulate the power flow from conventional sources, you burn more fuel (coal, gas, nuclear) than if you simply run them at their load factor. The ramp up costs fuel, whether coal or gas. You defeat the purpose of reducing conventional fuel use. Conventional power generation as stand by capacity is illogical. It is the exact opposite of the original design function of wind generation.

In fact, you pollute the local environment with NO2 and SO2: Nitrous oxide and sulphur dioxide. Review the footnoted pages 38 to 42[32] for a clear example of their noxious gas emissions. This is the notorious Bentek monograph from 2010. Wind energy sourcing is a clear and present danger to the environment.

Third, wind power sourcing fluctuates as power is sent into the grid. Recall that the grid has to maintain 85% of peak demand load to function. Backup power generation has to be maintained and constantly added to with each new turbine farm to compensate. While load factoring for wind-sourced power is a new and exciting field in power management, it is absurd. Imagine ballerinas with ten pound tutus and six inch nails through their heels. Applaud the dancing if you will, but it is ridiculous.

Fourth, assume the ideal wind generation figure of 20% of national peak demand is met by 2030. At a load factor of 25%, a massive amount of wind turbines will need to be installed—four times their nameplate capacity. Yet, the EIA feels that capacity factor in 2030 will be 40% for the U.S. system, in toto.[33] The only place where a 40% figure has been sustained is Tarurua, New Zealand, a rather wind swept region.[34] Ask the Kiwis about their birds and their own health issues.

Fifth, the fluctuations in power generation sourcing will easily out-power the 110% peak load demand, leading to *brownouts* as utilities preempt grid failure by shutting down demand via *force majeure*.

This has already happened in Spain and Texas. In Spain, the Union for the Coordination of Transmission of Energy tells the far scarier tale of winter 2006. The entire Southern European tier of nations, from Portugal to the Balkans, suffered a cascade failure. This was triggered by the switching off of a power line crossing a river. It tripped power, 40% of which was wind generated. The loss of the wind power further tripped the entire system into disfiguring itself into three autonomous regions. Transmission system operators were unaware that the wind power stations automatically reconnected themselves, preventing the three geographic regions from reconnecting to one another. The disaster was made worse by the lack of control at the wind farm source and the lack of awareness by the operators of these automated responses.[35]

Capacity losses, brownouts, and automated responses will become less endearing to the European population as more wind is brought on line.[36] These five challenges of physics will be exacerbated. They call it intermittency in Europe. It sounds like a disaster movie stuck in replay mode for years to come. An inconvenient Ground Hog Day. *Déjà vu* all over again.

Variability is the challenge.

Certainly, the wind industry has its professorial promoters. Most recently, at the 2014 AAAS national conference, Professors Jacobson, Bauer, Bazouin, et al, presented their storyline for 100% WWS (wind, water, and solar) energy sourcing for U.S. power demands by 2050.[37]

They propose 3.8 million 5 MW wind turbines, 1.7 billion new rooftop solar devices, 40,000 300 MW solar plants, 49,000 300 MW concentrated solar power plants (note their death knoll below, after one is brought on line), and a smaller number of tidal, geothermal, and hydroelectric facilities. Less than 1% of these facilities are currently in place. Their proposal would create five million new permanent U.S. construction jobs, 2.6 million operations jobs, while reducing energy costs by $3,400 per year and health costs by a further $3,100—all this while reducing air pollution deaths by 59,000 each year or 3.3% of GDP, and reducing by $730 billion annual global climate change costs. Their plan for the nation is bold.[38]

The only challenges they find confronting such a change in power production for the nation are upfront capital costs, transmissions needs, lobbying efforts by the fossil fuel industry, and political luddites from the right. They are quite right on all counts. These challenges are capital in nature: political and cash capital. Why they need to politicize the engineering masterpiece they describe is curious. If in fact they are right, then let the work begin. If their plans are smoke dreams or smoke screens, time will tell. The tale is certainly reminiscent of those windy stories from the 1980s, "full of sound and fury, signifying nothing."

One has only to cover a map of the U.S. with these structures to imagine the results. Who needs birds and bats, anyway?

Whether they are correct in their assumptions or on the merits of their arguments are for their peers to review. Their figures for capital costs are rather small, given the magnitude of the effort, something like $3 trillion. The source of these funds? You can guess.

As we are less interested in debating points over beer on a Friday night, let's go into some detail on the distribution and transmission issues.

CHAPTER FOUR
HARRIERS

https://vimeo.com/113068796

Watch the harrier's flight. It has terrific speed of chase. She then can hover in an instant when she sees prey. A quick drop from the sky for the kill. These birds are from New Zealand where they are fairly common in the wild. They enjoy small birds, mice, and road kill. The Maori call them Kerangi.

Their mating ritual begins in the winter with huge rises on thermals followed by crashing rocking dives. These circle plays in the sky derive their Latin name, circus, or circle. Their play continues during brooding, as the male brings food for the female, dropping it in the sky for her to catch. Their acrobatics are their life and joy. The chicks fledge in a few months, as spring approaches. The young birds first chase insects, catching them in the air as they practice for bigger meals. Larger than falcons, the harrier hunts from a glide while the falcon hunts from on high, diving to the prey.[39]

"When I run away to the circus I shall fly through the air on the trapeze." She meant every word of it. Little did she know that as a professional birder she would be with the circus. Studying harriers from the genus circus is her game. With 21 members the group permeates the globe. Population densities are decreasing, but not sufficiently to put any of the members on the Vulnerable List of threatened species.

Energy Distribution Technology

"Science is the belief in the ignorance of experts." — Richard Feynman

A wind farm may have dozens or hundreds of similar sized turbines connected via a medium size power collection system—the substation—and a separate communications system. The generators, substations, and capacitor banks required for wind farms need a constant supply of electricity to maintain their power factor responses.[40] These are rarely self-generated. Thus, the grid must provide the starting power for their energy generation. Once engaged at the proper wind velocity, the wind turbines spin, creating electricity. This is transmitted to the external system.

The grid takes the electricity. The task of the transmission system operator (TSO) is to manage both the security of the system (the grid) and the supply-demand flow. This may be easy to write in one short sentence. In the real world these heroes run the grid invisibly, nearly perfectly, all of the time across cities, regions and nations.

The power needs for a city such as New York can vary by 50%, or 2 billion watts, over a day, a season, a year. The region may demand twenty times that figure. Seasonal variances—summer sun, winter storms—are best visualized by massive ebb and flow tides. Imagine the entire Atlantic Basin filling and emptying every day. These TSO linemen regulate the flow of power through millions of miles of wire.

The wire is incredible. It is simple. Metals have the uncanny ability to channel electricity along pathways towards absolutely specific destinations: your iron or your IPad, your microwave or your light switch. The right wire acts like a pipeline, a conduit for energy. Power stations provide the pump action that forces the electricity down the wire. We consume the power.

The pipeline, the high transmission lines strung across and beneath America, transmits electrical energy at nine times the flow rate of gas pumped into your car at the filling station. A natural gas pipeline does the same thing at ten times the flow rate. A crude oil pipeline does so at one hundred times the flow rate.[41]

Metaphors are useful tools to explain a complex phenomenon. They are not, however, the phenomenon itself. In the real world of flowing electrons, massive wires and high flying electric cables thousands of miles long, someone has to do the actual work. When we said "the grid takes the electricity," we actually meant: the generator pushes the energy from the wind tower that has been converted from kinetic wind energy to electrical energy. This flows to substations where it is stepped up to high voltage. This is both a requirement for transit along the power lines and an increase in its efficiency of transfer. Higher voltage is like a wider pipe; it can flow more readily and efficiently.

Upon arrival at any substation down line, the flow is stepped down from the transmission level to the distribution level—lower voltage, or pipeline width. The power travels further to bus terminals where it is stepped down once again to fit into your home or garage.

The entirety of the electrical system is the grid. It can be radial, looped, or tiered ring in construction. Think of the branches of a tree or the blood vessels in your body. Now connect each of these lines—branches or capillaries—with at least two others. This is an attempt at redundancy. Why? Failure is common. Redundancy slows the progress of a failure, allowing the flow to pass along a different pathway.

This redundancy is vital. Generated power is consumed almost instantaneously (electric power cannot be stored except in small instances for short time periods: batteries do the trick poorly). Most electricity is generated by spinning turbines put in motion by heat. Heating coal, methane, steam or uranium nuclei creates power which rotates the turbines. The rotation converts the kinetic energy of revolving turbines into electrical energy which is supplied to the grid. In the case of wind, we have observed that the wind's kinetic energy is transformed into electrical energy through the rare earth permanent magnet generators and their gearing mechanisms. Different power sources flow through the same process and result in the same iron or IPad working smoothly.

TSO heroes and heroines manage the flow of power from generation through distribution to end user virtually in real time. Redundancy allows failures to be matched with flow from other sources, from other directions. It reroutes itself at the speed of light. When large scale failure happens, the entire city of New York goes dark, as it has several times in the past.[42]

When the flow of electricity is interrupted in a cascading series of events, many people are disadvantaged. This brings in the politicians. A fix is put up. Thus, a regulated industry, the utility industry, has more regulations imposed upon it. This industry is very capital intensive. Billions are spent on the massive components. These are built to last for generations. As systems evolve, previous ones become less useful, even obsolete. They must be replaced—more money and work.

It is also very *engineering intensive*. Highly educated professionals have to design and build the system. Other highly educated professionals have to run the system. These systems have evolved into vertically integrated structures, much like the governments that regulate them and the large corporations that can afford to profit from them. Horizontal integration is coming to the fore in the industry today, in both engineering design and operational management. Deregulation, when it happens, can require new approaches to construction management and to systems management. All of these temporal events alter the usefulness of capital intensive systems. Changes affect the cascade.

Alternative energy sourcing exemplifies all of these changes: massive capital deployment, multi-directional integration, and deregulation. Base load replacement by a regulated demand code promises to stretch a system already deployed in extremis. Intermittency challenges the operational system further.

Spinning reserves is the wind industry's answer to intermittency challenges. It is a euphemism for conventional power sources—coal, methane, and nuclear. Back up permits more wind sourcing. As more wind projects are built, second to second and minute to minute variability is accentuated by hour to hour minor variations.

Spinning reserves covers the problem. The reserves are ramped up, at some cost in power and CO_2 emissions.

What is the price paid for the efforts of TSOs to manage the intermittent wind energy flows? Accommodating the wind energy sourcing and its power generation concerns is costly to transmission functions—power lines, transformers, etc. According to the Edison Electric Institute (EEI), $61 billion likely will be spent on transmission projects from 2010 through 2021. 65% of this is being spent on "projects addressing the integration of renewable resources, and where needed, to accommodate the expected off-peak production." EEI adds that the needs of the new renewable generation that is coming online will require "the addition or upgrade of 11,400 miles of transmission" lines.

The cost for these lines has been estimated at $5 to $12/MWh by the DOE. The cost of the transmission lines needed to accommodate renewables will be borne by consumers.[43] That would be you, dear reader. Have a look at your electricity bill—costs for transmission are now stated on many of your bills, and these costs are rising.

In 2008 the DOE speculated in its *20 by 2030 Report* upon a 20% wind source for U.S. power demand. 300 GW of wind generation will be required, at a suspected cost of 50 cents per month for every U.S. household. This would avoid the need for 80 GW of coal plants, reduce CO_2 emissions by 825 million tons per year, and require substantial additional transmission construction.[44] Intermittency and spinning issues were left off table. Capital costs of decommissioning older power plants were ignored. Regulated demand pricing and multi-directional integration across the supply field were ignored. While I am also ignoring the details of these complex issues to tell a story, your government at work should have had the wherewithal to dig deeply into these systems design evolutionary pathways as it extrapolated its simplistic solutions.

The DOE had no idea of the impact the American shale gas fields were having on power sourcing, even as it was drafting the *20 by 2030 Report*. These fields have already reduced CO_2 emissions by more than 500 million tons since 2005 at a negative cost. As a nation we produce 10% less CO_2 than we did in 1992, the first year of measurement. We are doing so with an economy that has grown substantially in size and in CO_2 sourcing. It would appear that the future is resolving the concerns of the recent past. CO_2 emissions reduction is the baseline driver for the environmentalists. Are they satisfied with this massive reduction in the problem they describe so nightmarishly? Well, no. Apparently the problem is even more hellish today than ever. The Administration's EPA has just announced its demands for CO_2 reductions by 2050. It takes as its baseline 2014, conveniently ignoring the 10% CO_2 reduction already achieved. It expects a further 30% reduction. Guess who pays for this delivery from hydrocarbon Hell.

Here we must bring in another euphemism: reduced expenses for electricity for consumers. No tax credits, no regulatory demand, no preferential treatment of power source, the 825M ton CO_2 reduction objective of the DOE report should be achieved entirely through domestic shale gas production and distribution within another five years. Is this counted in their figures? Hardly. It would skew the results to methane, not exactly in their deck of cards.

By 2030, we can only speculate as to the reduced CO_2 emissions from domestic conventional gas energy production and sourcing. U.S. based production of natural gas has increased every year since 2008—by 47% in 2014. Use of dual cycle natural gas turbines in the newest utility power generation stations has several consequences: reduced cost of power generation, reduced CO_2 emissions as coal is supplanted, reduced dependence upon foreign fuel sources, and more jobs.

You could reasonably conclude a 1 trillion ton reduction in CO_2 emissions by 2030 from conventional power generation. Who knew?

Or you could demand alt-en power sourcing to achieve a lesser result at far higher costs. Expenses for consumers of electricity—business, farm, and consumer— would skyrocket as more conventional power generation is required to cover the hides of the alt-en crowd and their inefficient power sourcing.

Let's allow the wind industry its word on costs. A March 2015 report[45] from DBL Investors of San Francisco is written by the Managing Partner Nancy Pfund and her associate Anand Chhabra, a lawyer in training from Stanford. In full disclosure she acknowledges that she sits on the Boards of several alt-en companies including Solar City. The paper, *Renewables Are Driving Up Electricity Prices: Wait, What?*, thanks the NREL, an MIT professor, the Berkeley National Laboratory, and the staff of Solar City for their support. Their conclusions: "greater generation of electricity from renewables did not mean skyrocketing electricity prices and states with renewable electric power generation often experienced average retail prices well below states producing less electricity from renewables."

Let's try to understand their meaning, in all fairness. They illustrate price increases as a percentage, then compare these percentage increases across all states, as well as between the 10 with the highest renewable sourcing and the 10 with the lowest renewable sourcing. The price increases as a percentage is

renewables leaders:	3.08%/year
nationwide:	3.23%/year
renewable laggards:	3.74%/year

The data source is the EIA, our friends at the Energy Information Agency, no laggards when it comes to data generation. The point the authors are making? The difference in unit cost of renewable electricity has changed from being $.0015/KWh (that's fifteen one hundreds of a penny) more costly to $.0049/KWh less costly. According to this differential, if you use 100 KWh each month and move from California to North Carolina, you electricity bill will drop by 64 cents a month. If you use 1,000 KWh each month, your bill will drop by $6.40 each month. I can attest to the fact that our electricity bill changed once we moved to North Carolina from California. We use slightly more during the summer and slightly less during the winter. We use about 10% more electricity overall. The bill in North Carolina averages $28. The bill in Ojai, California averaged $89. Anecdotal data is useless to a statistician, certainly. The bill remains lower, nevertheless. Nancy Pfund is welcome to visit or review my bills.

Concerning jobs, we have this information from a wind energy source, the irrefutable American Wind Energy Association (AWEA), and a wind farm owner, Jacob Susman of Manhattan, New York. He wrote in August 2015[46] that wind energy has created and sustained 73,000 jobs across America. His source file is an AWEA report (accessible for only $550.00 or by paid subscription). Unwilling to spend this sum, I am reliant upon Jacob's veracity. His own firm, OwnEnergy, has built out farms with 289MW of installed capacity (that would be 81 MW active capacity at 28% usage) that produced 880 construction jobs with about 12 permanent jobs. His footnote indicates that this information is not from his payroll department but "from the NREL Jedi Wind Model." He further states that the CO2 offset is 751,400 tons each year from his 289 MW turbine farms. This too is from the model. He comments: "No wonder wind power is an industry that receives bipartisan support."

What a Jedi Warrior! Almost 12 permanent jobs (does that mean someone is working overtime?). Incredible. His farm is in Texas.

So let's do the numbers!

He sells at the guaranteed price of 2.5 cents/KWh.

289 MW x .28 delivered capacity = 81 MW average generated capacity.

1 MW = 1,000 KW hourly generation.

81,000 KW x 2.5 cents/KWh = $2,025/hour.

Of course, this is his floor, his guarantee. As demand peaks, the price does as well. The sky's the limit. To the universe and beyond!

The *20 by 2030 Report* concludes that transmission systems could and would be built. The only item missing from this agenda is the cost factor. The new EPA demands for clean energy sourcing also ignores these quadratics.

Let's dig a little deeper into the power demand and supply equation. *Distributed generation* and *demand response* are new phrases that may enlighten the nation in the near future. We have focused entirely upon the economics of supply. If we examine demand elements in detail, we may realize additional means of energy consumption patterns that actually flow backwards to the supply side of our equation.

Distributed generation may add to the TSO's systems challenges. It may also reduce these challenges. No one today seems to know. This is the substitution of many small power generators for fewer, larger generation sources. The idea is laudable. From the capital, systems, distribution and efficiency point of view, thousands of smaller power sources cost less per unit. They offer multiple magnitudes of redundancy and use shorter, thinner wires. They act more efficiently over lesser distances. The small wind turbines we discuss later are nearly perfect examples of distributed technology applied: many homes could have one of these small VAWT turbines on the roof to augment local, conventional power supply.

The difficulty lies in the implementation. As new buildings have these small systems built in, they will require less grid power sourcing. If this cascades across a region, such as is happening in Arizona today, the grid operators may cry foul and require a higher fee for service to offset lower demand. They may charge additional fees for short line access through their grid system. They may simply fall into bankruptcy, causing a capital cascade of different proportions. Will the federal government step in to save the system it has tried to eliminate? Will capital flight reduce distributed power to smaller locales?

On the other hand, the fear of grid attacks from enemies real and supposed may be thwarted by the design of the system. Let a thousand lights—and all that.

Distributed generation may allow the return of DC power, igniting the century old debate between Tesla and Edison. There are no physics reasons for either AC or DC to be preferred in a delivery system, other than the internal structure of the system itself. Today's systems are designed for AC, as per instructions from Mr. Edison. Mr. Tesla could work his alternative magic in a distributed system world. Regional and national grid design would take on new dimensions.

Demand response allows the TSOs to monitor and restrict electric power use in the home or business. By reducing excess demand during peak usage, this new system management tool offers greater flexibility in dealing with the major challenge of alternative energy—intermittency. Solar has problems similar in scale and dimension as wind. If demand can be controlled, the intermittency issue may be muted. Three approaches to DR are discussed today: emergency, economic and voluntary.

Emergency response requires the utility company to have direct control over various components of your home's electrical devices. In our home, we have agreed to have a governor installed on our AC systems that can reduce or eliminate our demand. It can take over our AC. In return, we are offered an *economic incentive*: lower electric fees. This in turn conditions us to a *voluntary response*—social conditioning—to further requests for device controls. All three of these are key to the demand response system approach to the horizontally integrated power sourcing in today's world.

Taking these ideas to another level requires the consumer to willfully act in regards to energy use. Smart grid application encourages such behavior. We all aspire to a wise use of insulation, up-to-date electrical devices, and acceptance of responsibility of each family's energy needs. We may soon accept incentives for energy use at all times. We also know that each of us is an economic beast. We respond to fair prices. If the price for our energy use is time based rather than fixed unit based, we may quickly adapt energy management systems in our home. These smart meters give us better awareness and imply better control of our energy demands. If you are incentivized to be smart—if you pay less—then you win. The ancillary winner may be the economy and the environment. Applied technology encourages further technology drive.

Businesses already are smart about monitoring, shifting, and balancing energy use across time and space. What functions can be performed at off-peak times in non-urban buildings and how can they be monitored for minimal power drain?

How does this free up time, personnel, and capital that can be more efficiently employed to drive more sales at lower prices to our eager customers? We and they earn increased capital flows.

The demand side of energy is under exploited. Few have taken advantage of its incentives. Because of the distributed nature of multiple utility companies, a wide variety of experiments can be explored. The federal system of government was specifically designed to encourage experiments in governance at the state level. These firms can similarly explore joint opportunities between themselves and their customers.

Full implementation across the coverage area of the 500 utility companies in America may take time. It may be imposed from above by a federal regulatory engrossment, completely disfiguring the market-based capitalist system we have chosen. In fear of climate change, anything may be wrought by committed commissars.

These regulatory soviets may dictate power use to the people. The data flows both ways, by the way. If it is easy to read your power use, it is equally easy to see which source of power drain you are using and how efficient it is. It is also simple to see the clicks you use on your keyboard, just as the links you visit on the internet. How far are you willing to have Mother State delve into your personal electronic usage? How much choice do you have today? Will you have any choice in the near future?

Top down management has never proved successful at any scale other than the smallest. It is proving unsuccessful today in the alternative energy industry. Rent seeking replaces capital seeking. Taxes taken from the citizenry are rarely deployed to advantage by a government. Even at their best, they impose a cost. No tax dollar generates more revenue than it would in a private arena. Multiplier effects are nightmares dreamed by the insane or the overzealous in fear of an invisible future. Prediction invariably leads to predation.

Wind Energy Application

Here is a current list of the largest projects in the U S. as of 2011:

Locale	Installed Capacity	# of Turbines/Manufacturer
Altamont, CA	981 MW	7,377 / Vestas

Roscoe, TX	782 MW	627 / various
Horse Hollow, TX	736 MW	421 / GE, Siemens
Capricorn, TX	662.5 MW	407 / Mitsubishi
Sweetwater, TX	585.3 MW	392 / various

Source: *AWEA, Annual Market Report, 2011*

There have been 3,464 turbines installed in 2011, 2,942 turbines installed in 2010, and 5,765 turbines installed in 2009. As of 2012, there are more than 40,000 turbines with 52,000 MW of installed capacity in the U.S.[47] This figure has risen to 60,000 MW in 2013. Installations of less than 100 kW are defined as small wind turbines.

Offshore facilities have yet to appear here in America, although they are covering the Baltic and North Sea regions of England, Denmark, and Germany. A facility has received federal lease approval near Nantucket, Massachusetts and fifteen other sites are awaiting regulatory review. The aesthetics are an issue for a few friends on Cape Cod. It seems they'd love to use wind power as long as they don't have to see it from their enormous backyards or their play yachts.

2012 saw the introduction of 18 GW of gross wind power installed capacity, 10 GW of solar, and 55 GW for natural gas. Note these figures are *installed capacity*.

Let's do the math: Recall the distinction between this gross figure for wind and the net, or 25 to 30% (40% under ideal conditions). This is the *capacity factor*. The comparable figure for natural gas is 85 to 90%. Thus, the usable, distributed energy figures now read, natural gas generated energy 49.5 GW vs. 6 GW for wind generated energy.

	Gross capacity	Distributed energy
Wind	18 GW	4.5 GW
Natural gas	55 GW	49.5 GW

One case may suffice to illustrate. On February 26, 2008, the wind energy dropped 75% (1,500 MW) within three hours in west Texas. This happened during afternoon rush hour for the urban cowboys of 21st century Tejas. By 6:30 pm the AC running through the transmission lines was sluggish. System wide collapse was imminent. Previously agreed upon power cuts (brownouts) were applied to many industrial plants. The reduced demand of 1,200 MW allowed the lines to survive for another battle. The system stabilized.[48] The lesson? Integrating wind energy into a grid is difficult and will only get more difficult as the wind energy multiple increases.

In the United Kingdom, a recent study[49] has revealed that monitored wind turbine output (as measured by the National Grid) increased from 5,894 MW to 8,403 MW over the period.

- The average capacity factor of all monitored wind turbines, onshore and offshore, across the whole of the U.K., was 29.4% in 2013 and 28.8% in 2014.

- The monthly average capacity factor varied from 11.1% (June 2014) to 48.8% (February 2014).

- The time during which the wind turbines produced less than 10% of their rated capacity totaled 3,278 hours or 136.6 days over the two-year period.

- The time during which the wind turbines produced less than 5% of their rated capacity totaled 1,172 hours or 48.8 days over the same period.

- Minimum wind turbine outputs averaged 132 MW (1.8% of capacity) in 2013 and 174 MW (2.1%) in 2014 as measured over 30 minute intervals.

- Variations in output of 75 to 1 have been observed in a single month.

- Maximum rise and fall in output over a one hour period was about 1000 MW at the end of 2014 with a trend increase of about 250 MW per year as measured over four years.

- There is no correlation between U.K. wind turbine output and total U.K. electricity demand, with output often falling as demand rises and vice-versa.

- The conclusions to be drawn from the analysis are that the increase in nominal capacity:

- Does not increase the average wind turbine capacity factor.

- Does not reduce the periods of low (less than 10% of installed capacity) or very low (less than 5%) output.

- Does not reduce intermittency as measured by average monthly minimum output.

- Does not reduce intermittency or variability as measured by maximum rise and fall in output over a one-hour period.

- Does not indicate any possibility of closing any conventional fossil-fuel power stations as there is no correlation between variations in output from wind turbines and demand on the Grid.

Therefore, based on the above, there is no case for a continued increase in the number of wind turbines connected to the Grid or for the associated subsidies for wind energy, since this is an ineffective route to lower carbon dioxide emissions.

This summary is devastating to the British wind energy industry. Variations, variability, and wind strength dependencies are clear. It helps to explain why the national government has taken deep and abiding second looks at further expansion of the wind energy grid. Local communities are more successful in determining their future land use, tax issues, and power generation sourcing.

Since 1998 in the U.S., average turbine nameplate capacity has increased by 170%, average hub height has increased by 50%, and average rotor diameter has increased by 96%. These design characteristics are employed to advantage lower wind speeds.[50] Height is intended to bring the blades up above ground interferences, to the steadier flow of air space. Hub size increases both allow for slower speeds and more generated power. Height also impacts seasonal flight patterns of nocturnal avian populations as well as migratory windglider populations. Weather concerns—cloud cover, mist, rain, and turbulence or lack thereof—can drive birds higher to seek easier migration or better hunting.

We shall investigate alternative wind tower and blade designs later. These FrankenTowers have reached the limits of their design perfectibility. They have not reached the limits of bird mortality.

Have we been told the truth? Have we been told that nameplate capacity is far less than the capacity factor? The difference is enormous. The less than gentle urging to build yet more towers disguises their extraordinary inefficiencies. The simple difference between nameplate capacity and net capacity is 13.5 MW in our example. Under the right conditions, operators can reasonably expect to get a load factor of 25-30% of nameplate capacity. This is a hidden truth. What we are told by the industry is an outright lie. This is another of our many disingenuities. Broken promises yet again.

Europe

The European Union is keen on wind power. They have sited much of it offshore, at enormous expense.

The most recent disaster to fall upon the European wind industry is the Atlantic Array. The Bristol Channel has been the starting point for many of England's voyages of discovery and conquest. Its 43 foot tidal range is second only to the Bay of Fundy. Wading is a local pastime when the tide is out. The Bristol Channel is also viewed as the most important wind alley in England. If properly harnessed, it could provide 5% of the U.K.'s electricity needs. The Atlantic Array was to be the centerpiece.

As many as 220 monopole turbines were planned, each as tall as 720 feet, connected to four offshore substations via eight subsea cables. The name plate capacity was to be 1,200 MW (capacity factor at 28% = 345.6 MW). The cost? £1.5 billion. It hoped to reduce CO_2 emissions by as much as 452,000 tons each year. Let's do the math. At the current exchange rate of $1.65, the U.S. dollar cost is $2.5 billion. That comes to about $2M/MW or $2,000/watt of electricity

Now that's a pound of flesh. In November of 2013, the entire project was canceled for economic reasons. No? Really? Locals were delighted that their efforts to Slay the Array were successful. U.K. Green gobs were unsettled with the statement, as they felt the cost of wind was coming down... investment in English alternative energy schemes has dried up by the end of 2013. Even the coalition government stepped back from the projected £1.5 billion investment cost.

CHAPTER FIVE
SEA EAGLE

Sea eagles are what their name suggests: they fish from the sea. In addition to fish, they enjoy frogs, sea snakes, turtles, crabs, and even bats. Sea eagles are quite large with a wingspan of 60 to 70 cm. They climb on thermals for vantage, then skim the surface and take their prey with a backward slash of their talons. Mating for life, they tend to maintain large territories for food and game hunting. The images here are from Singapore where these massive fishers live in mangrove swamps. Burung Hamba Siput means shellfisher, as they appear to love shellfish. They build large nests in high trees. The mother lays two eggs which she incubates for several weeks. The male provides food and protection until the chicks molt.[51]

Of the 248 species globally, several are endangered. Turbines present a threat to these birds when sited offshore or near swamps and low water estuaries.

https://player.vimeo.com/video/120049225

Ill blows the wind that profits nobody. — *Henry VI*, Part III, Act II,
Shakespeare

Economics

How do the economics of supply, demand, and pricing affect the wind industry?

How does legislative, regulatory and legal action affect this nascent industry and local society?

What are the differing approaches taken is various countries: Spain, Germany, Great Britain, and America?

According to the American Wind Energy Association (AWEA), "adding wind power saves consumers money and helps hold down the costs of other fuels." AWEA is the lobbying arm of the wind energy industry. These are neither scientists nor academics; they are rent seeking capitalists, just like the American Petroleum Institute (API). The folks from AWEA have a four pillar agenda, implemented in Washington, DC by their lobbying efforts, their PAC and their extensive relationships with members of Congress. These four pillars are:

- Drive demand for more wind energy.

- Make wind energy as cost competitive as possible.

- Implement policies to help deployment and operations.

- Build political strength.

This organization, like any political force, seeks to encourage laws and regulatory framework conducive to further business expansion. All lobbyists do the same. They are not evil capitalists; they are simply capitalists, very powerful political capitalists.

How long have wind farms been in operation? The history of wind use virtually predates written communication for mankind. Small vessels have been wind-driven for tens of thousands of years. Grinding of seed has been in use since at least 3000 BC. Wind-driven power sourcing has been recorded in the U.S. as early as the 1880s.

1975 through 2002 saw the first federally subsidized wind farms. By 2002 there were enough in the nation, 149 farms, to provide power for 1.1 million homes. By 2012, there were 815 farms supplying the energy needs of 15 million homes. A decade of development resulted in a 15-fold increase in power potential.

Installation capital costs have ranged from $700/kW to $1,400/kW between 2000 and 2012.[52] While wind turbine/tower lifetime is estimated at 20+ years, reliability and performance are impacted by turbine failures, in part or in toto. The equipment is expected to work, virtually unattended, for two decades.[53] Note that O&M contracts have grown in number and price over the past decade as manufacturers are profitizing the long term nature of their build. The prior reference makes the interesting metaphor of these contracts being carbon copies, certainly without prejudice!

When the equipment does not work, is offline, or fails structurally, productivity declines. The declines are a result of a wide variety of factors: siting, wind strength, season, diurnal wind flows, size, capacity, connectivity, demand, regulations, and legislation. These declines may or may not be built into the economics of wind. In the U.K. during 2013-2014, the capacity factor for the nation was 29.8%, while varying hour to hour by as much as 75 to 1. Demand had no correlation to supply. Hourly intermittency was as much as 1,000 MW.[54] This massive variability in both available power and durability of said power ensures that wind has no chance of replacing conventional power sources, ever. Period. It is impossible.

Installation of 70-ton turbines and 150-foot blades requires enormous expertise, planning and logistics, not the least of which is wind speed on installation day. Plants are concentrated in this patchwork of states: Kansas, Texas, Iowa, Colorado, and Arkansas. Similar plant sites are located across Europe. To minimize transportation costs and delivery schedules, much of the installation is nearby. As much as 67% of production may happen onshore, although this figure is malleable.[55] Imports have declined as local production has ramped up.

In one sense, the industry is new and growing. But in another sense, it is 40 years old, geriatric, fixed in its ways, inflexible, and demanding. Any industry experiences boom/bust cycles. The wind energy industry seems to exacerbate these cycles with massive capital and tax flows that either buoy or submerge the participants.

Energy so based upon rent seeking will always be extravagant, excessive and extreme. There are 82 different wind energy subsidies overseen by nine federal agencies. If this ain't free money, what is? This is taxing the poor to feed the wealthy. Don't take it from me: Warren Buffett, who has invested billions in renewable energy, stated,

> We get a tax credit if we build a lot of wind farms. That's the only reason to build them. They don't make sense without the tax credit.[56]

Don't want to listen to the third richest man in the world? How about the Richest One? Retired software kingpin and richest man in the world Bill Gates has given his opinion that today's renewable-energy technologies aren't a viable solution for reducing CO_2 levels, and governments should divert their green subsidies into R&D aimed at better answers.

Gates expressed his views in an interview given to the *Financial Times*, saying that the cost of using current renewables such as solar panels and wind farms to produce all or most power would be "beyond astronomical. The only way you can get to the very positive scenario is by great innovation. Innovation really does bend the curve."[57]

Subsidies are their only reason for investing. The result? According to a recent study, wind energy subsidies force the cost of wind sourcing up by 48% from already high levels for 2015.[58]

In terms of sources of current power generation, here is the table for 2010 (and 2012 in parentheses) from the Department of Energy (DOE):

Coal	45% (37%)	1,851 GW
Natural Gas	25% (30%)	1,030 GW
Nuclear	19.6%	807 GW
Hydro	6%	257 GW
Wind	2.3% (3.5%)	95 GW
Solar	2.5%	91 GW

The table has changed remarkably over the past decade. Wind has tripled it sourcing, from 1.2%. Natural gas has risen from less than 2%. Nuclear and hydro remain constant. Solar has finally made the scorecard. As a whole, inclusive of hydro, renewable sourcing for U.S. electricity production has risen from 9.4% in 2000 to 10.3% in 2010 and 12% in 2012. Many would exclude hydroelectric power from the list of renewable energy sources. If we do so, renewables have risen from 2.1% in 2000 to 4.8% in 2010 and 6% for 2012.

In terms of federal capital funds (tax credits, subsidies, and support), the DOE states the following through its EIA, or Energy Information Agency, one of the best information gathering locales in the federal government.

Fuel	Share of Subsidies/Support for Fiscal Year 2010
Coal	10.0%
Natural Gas	5.5%
Nuclear	21.0%
Renewables	55.3%
Biomass	1.0%
Geothermal	1.7%
Hydropower	1.8%
Solar	8.2%
Wind	42.0%
Transmission	8.2%

These subsidies are made up of five classes of support:

> Direct to producers/consumers
> Reduction in tax receipts
> R&D
> Loans and loan guarantees
> Geographically-targeted electricity programs

These classes exclude depreciation, municipal bonds, foreign tax credits, PTPs, Section 199 deductions, Ex-Im Bank tax credits, energy-related trust funds, and nuclear liability subsidies.

The subsidies are regarded in the federal budget as *revenue losses*. Such losses totaled $16.3B in fiscal year 2010. The funds are paid for from eight of the twelve departments of the U.S. government: Energy, Interior, Labor, Transportation, HUD, HHS, Agriculture, and Treasury.

Coal	$0.5 billion	3.5%
Renewables	$8.3 billion	50.3%
Hydrocarbons	$2.7 billion	16.5%
Nuclear	$0.9 billion	5.5%
Consumer/producer energy efficiency	$3.9 billion	24.2%
Total	$16.3 billion[59]	

According to the DOE[60] the following table represents its estimate of levelized cost of energy, expressed in U.S. dollars per megawatt hour:

Plant Type	Levelized Cost of Energy (USD/MWh) – Median
Wind, onshore	60
Wind, offshore	100
Solar PV	280
Solar CSP	200
Geothermal Hydrothermal	60
Enhanced Geothermal	130
Hydropower	20
Ocean	220
Biopower	70
Distributed Generation	140
Fuel Cell	150
Natural Gas Combined Cycle	50
Natural Gas Combustion Turbine	70
Coal, pulverized and scrubbed	50
Nuclear	60

Combined cycle natural gas is the star on the list, (with its older cousin combustion turbine a close rival), even as it is buried in the data pile. Hydropower and geothermal are in second place, as expected. Coal settles into third, in a dead heat with nuclear and onshore wind. Solar brings up the rear.

These figures evolve over time. Combined cycle natural gas is a relative newcomer. Geothermal and hydropower have been here for decades. Wind has been on the list since 1992. Solar has been eating dust since the 1980s. Nuclear has experienced a remarkable fall from power, as costs associated with decommissioning plants continue to rise.

The evolutionary nature of these figures should give pause to any prognosticators. Like an unguided flock of birds or a wandering school of fish, the energy industry has few overriding principles. Complexity Theory may be the only one to withstand time's erosion. These energy sources as a fixture in the U.S. economy are as changeable as the birds—and the Tax Code.

The wind, like the sea and the desert, is inscrutable. Very few physics principles,

clearly stated, change over time. The few remaining concepts are relative in importance and in application. Their importance today may be reflected by prejudices
of long ago. The application of relativistic formulas from our past to the present
is usually informed by prejudices. These prejudices can be ideological, cultural,
economic, or purely personal. When these preconceived prejudices are introduced,
much of the world responds slowly. The removal of these relativistic formulas takes
much longer.

Meanwhile, rent seeking capitalists seek to control the gate and the gatekeeper.
This tends to be the feral coffers, the federal fisc. All know that once a gate is open,
it can more easily be opened further than shut even a tiny amount. Power to write
regulations is the power of right legislation. Promulgating fear opens doors. This
fear story is best told by distraction. Distract the commons with fear, write laws
that encourage growth in your industry, take the money and move on.

The energy industry tends to move to new rhythms at a glacial pace. The
imposition of these new ideologies (from greed to climate change) on the flock has
unintended consequences that ripple through the economic fabric of the industry.

Take Ontario, Canada for an example. This is from a recent report titled
Ontario's Electricity Dilemma by the two Ontario engineering societies, the Ontario
Society of Professional Engineers (OSPE) and the Professional Engineers of Ontario
(PEO):[61]

> *The province is in fact substituting electricity that produces an average of
> 40 kg CO2 per megawatt-hour (from gas turbines operating ONLY during
> peak demand) with electricity that produces an average of 200 kg CO2 per
> megawatt-hour (from gas turbines that MUST operate whenever the wind
> stops blowing).*

Local participants and national players alike are often caught out in a storm
of change. Venturing forth is always a risk. Planning a voyage is 90% of the effort.
Man is less the master of his fate than a voyager across an unknown sea. It is
dangerous to change the rules of the game. Better to follow in step, pick up scraps
and banquets as you may.

The DOE likes to rummage about in the future. The impossible nature of any
speculation about the future does not seem to hold back our federal masters. What
does this DOE see this table to be in 2018?

Estimated Levelized Cost of New Generation Resources, 2018

U.S. Average Levelized Cost for Plants Entering Service in 2018

Plant Type	(2011 USD/MWh) Levelized Cost
Conventional Coal	65.7
Advanced Coal	84.4
Advanced Coal with CCS	88.4
Natural Gas Fired	
Conventional Combined Cycle	15.8
Advanced Combined Cycle	17.4
Advanced CC with CCS	34.0
Conventional Combustion Turbine	44.2
Advanced Combustion Turbine	30.4
Advanced Nuclear	83.4
Geothermal	76.2
Biomass	53.2
Wind	70.3
Wind – Offshore	193.4
Solar PV	130.4
Solar Thermal	214.2
Hydro	78.1

The table is slightly different, certainly. The order has changed and the figures are quite different. You will also notice whole new additions to the technology for sourcing power generation: "advanced" being the applicable word. Without going into the details of these new distinctions, we can see the projections are each significantly lower, in terms of 2011 cost. The biggest winner is natural gas power generation with a decline in cost from $70/MWh to $16-18/MWh. That is an 80% drop in cost. Wind comes in at 58% decline. Solar finally joins the fleet, with a 78% drop in power generation cost. Nice work, if you can get it.

These ideological projections of present beliefs for our future nation are, at best, guesses. They reflect the current state of political energy affairs in Washington, DC. As such, they are also a mirror into the soul of the political beast. This house of cards may last, may be built upon a sea of sand, or may have no bearing on the future. As guesswork, it is interesting. Take it for just that and nothing more.

LCOE,[62] or levelized cost of energy, is as important an acronym in the wind energy industry as EUR (estimated ultimate recovery) is in the hydrocarbon energy business. It is an attempted mathematical expression of all variables in the production of wind energy.[63] The cost is expected to drop annually by 1 to 6% between 2013 and 2030. This is a potential total cost reduction of 20 to 30%. For comparison, the costs dropped 65% between 1985 and 2004. The timeline is roughly similar, implying a constant gradient of improvement in technology and capital investment.

Note that these costs are unsubsidized; that is, without tax support at either the state or federal level. The tax event is thus ignored when NREL does the calculations. Any subsidies, tax credits, et al, are outside of these schema. These calculations are thus very honest. They are asking the question: Can wind stand on its own economic feet? The tax event adds value to the invested capital as a subset of return on equity—one that is guaranteed by Congress. In the end, the economics of wind energy production must include tax subsidies. These are plentiful and fruitful. Virtually all flow to the investor of capital. Some flow to certain end-users.

Let's examine one deal in detail[64] as an example of these economics in action. Google proudly announced on September 13, 2013 that it had entered into a contract to buy all of the power generated for the next twenty years by the American Indian owned Happy Herford Wind Farm in Amarillo, Texas. The company's stated environmental goal is a zero emissions policy. One of their avenues to achieve this policy is wind power. The vehicles running down this five lane freeway are PPAs, RECs, PTCs, accelerated depreciation, and RPS.

In baseball, three pitches for strikes and the batter is out. In this game, you must decide who the batter is. He will get five swings at the ball.

Power Purchase Agreements (PPAs) define the wholesale market price for wind power electricity. They are the contract between two parties, the firm that generates electricity and the one that purchases electricity. The PPA defines all of the commercial terms for the sale of electricity between the two parties. This includes timing of the project, delivery schedule of the power, penalties for poor delivery, payment terms, and termination. It is the primary agreement that defines the revenue and credit quality of a project and is instrumental in obtaining government favored, tax sponsored project finance. From 2003 through 2008 PPA pricing compared favorably with other sourced electricity. In 2009, natural gas sourced from fracking in the Marcellus began to impact an economy already in a recession. Prices plummeted for wholesale electricity.

Through 2012, wind source electricity was entirely priced out of the market. In other words, it cost far more than any other power source. You may have noticed this on your electricity bills over the past six years. The wind energy generating farms receive the production tax credit, soon to be revealed. This guarantees them a price for their product and a market that must buy it.

Your power company still has to buy the electricity from wind generators. They simply pay top dollar for it; well, you the consumer, pay top dollar for it. By the end of 2012, wholesale electricity prices at the very top end, rose sufficiently to at least begin to include wind power sourced electricity.[65] Lucky you.

PPAs are the vehicle of choice for the Happy Herford. Google's contract is a PPA. They will buy all the electricity produced for 20 years. Good work, Google. As an example from the petroleum world, Cheniere Energy, the new LNG natural gas exporter whose four-train terminal opens November of 2015, has already sold 80% of its capacity for twenty years—FOB at delivered market price plus 15%. You would love to have a 20-year buyer for your product or service. Both Cheniere and Happy Herford are the winners.

Will Google use all this pre-priced guaranteed delivery electricity? No. As they indicate, sites for data storage and sites for wind turbines have different characteristics. Rarely do they match. Too much *open space architecture* between source and user. The company, in its wisdom, is buying the wind power, then immediately selling it in the open market for energy in north Texas. It will continue to purchase all of its electricity needs from local power suppliers in its various data center marketplaces.

Here comes the fast pitch on the inside corner. Strike One!

Renewable electricity credits (RECs) have been issued. These *societal benefits* have been dictated by the federal government as an economic tool, both to encourage the move to green energy production and to reduce the use of so-called polluting fossil fuels. The tools may be used at the consumer level with *energy rebates*. But they are almost exclusively used at the commercial level to earn economic benefit from the production of green energy, as defined by the EPA. These benefits are not a measure of electricity. They represent the deemed social result of the effort to make it. They are a commodity and are traded as such. When purchased, the new owner can now use them as a claim against their conventional energy use—they are an offset. One might be tempted to call them an indulgence.

No reduction in conventional energy consumption has actually occurred by Google. They are still a sinner. But their credit has extinguished a debit and they are free to go. Thus, Google can say without impunity, "One REC is created when a MW of green energy is generated and one REC is consumed when a user says they have used a MW of green energy."[66]

Curve ball, inside corner. Swing and a miss. Strike two!

Casey doesn't know it yet but the developers of Happy Herford (such a nice cow) will receive a lovely *Production Tax Credit* (PTC) of 2.6 cents for every KWh of electricity they produce, for ten years, adjusted for inflation.

Let's do some math:

The nameplate, or installed, capacity of the farm is 239.2 MW (239,200 kW).

The capacity factor in the local area is 45%.

There are 8,760 hours in a year.
Annual energy production should be 942,926,400 kWh.[67]
This figure x 2.6 cents equals $24,516,086.40.

Over ten years of tax credits, the happy cow has earned $245,160,864.00.

This is real money. A tax credit is the best deal to receive from the federal government. It is a dollar for dollar reduction in tax liability. It is a fungible commodity that acts as a currency. It can be traded at a price dictated by the marketplace. Nice work.

Knuckle ball inside. Strike Three.

Poor Casey. The umpires have added two more pitches to the game: accelerated depreciation and avoided renewable portfolio standards. The *accelerated depreciation* deduction allows the developer to deduct all of its capital costs over the first five years. Most companies in most industries get to use some form of accelerated depreciation. In the oil business it is called IDC, intangible drilling credits. In real estate, the time frame is 26.5 years. In your business you have a variety of depreciation schedules for desk, computers, cars, plant and office equipment. The amount depreciated and the time you can use for it differ. No one gets a five year schedule any longer—except for your auto. And desk gear. Small beer in the depreciation world.

Google gets five years for a massive depreciation amount. It has first exchanged an intermittent energy source (wind) for a perfectly reliable one (conventional fuels). Second, it buys at a preferential rate and immediately sells its contract as RECs. It has also wisely hedged its future cost of energy. Third, it is now invulnerable to any future proposed *Renewable Portfolio Standards* (RPS). These are pesky political gizmos that define energy, renewable, and standards in a whole new light. These RPSs may be used against large power consumers. RPS negatively impacts competitors and now have greater societal importance to Google. Fourth, the environmental PR is an equity kicker for the shareholders, most of who seem to be concerned about the future of the planet's ecosystem. Finally, the stock price may rise as equally concerned citizens bid to acquire such a valuable security. Homerun for the good guys—who weren't even at bat.

Casey strikes out.

So, you ask, who just struck out? It seems that everyone wins in this game: the pitcher, the players, and the fans.

You, the taxpayer just got hosed, for $245 million in greenbacks. Too bad about the late bill. Your tax dollars at work, you see. The feds pay an ancient debt to the remnants of an Amerindian tribe for 19th century atrocities. A major global corporation earns PR and cheap energy and tax free income for years. The government has shown the way to the future. Consumers get to have their web work watched secretly by uber wealthy super sleuths—and your kids get to pay the tax bill. Nearly a quarter of a billion dollar deal goes down on Friday the 13th. Check the footnote, sailor. It's better for the environment, better for your children, better for the future of the planet. Kumbaya, friend.

2014 saw the removal of wind production tax credits, or PTCs, in the last days of the previous year. Congress wrapped up its tax planning session with an end run around the wind energy industry. It won't take long for this credit to re-emerge in the Tax Code. As we write, the credit is reentering main stream taxland yet again. The recently signed Omnibus Tax Bill of 2015 has extended wind-based tax credits for five more years.

These billions, $12 billion plus of taxpayer dollars, have helped support the emerging energy industry for more than forty years. The subsidy, once in place, lasts for ten years. It is put in place at the start of a new project, when 5% or more of capital is committed. This allows the wind energy producer to sell energy at a premium price ($.022/kWh today), while potentially producing it at a loss, over the following decade. Profits ensured, subsidized by the taxpayer and/or rate payer.

Tired of the endless beating? Shall we simply ignore the inequality of capital cost/KWh between the wind industry and the nuclear industry?[68]

How do costs compare for power generation between wind, coal, natural gas and nuclear? If you integrate all costs of design, construction, O&M, and decommissioning and compare new to existing facilities, you arrive at this column from the Institute for Energy Research. Yes, it is a subset of The Heartland Institute. We are trying to be balanced in our quotes.

Coal	$38.40/MWh
Natural Gas	$38.90/MWh
Nuclear	$29.60/MWh
Wind	$112.80/MWh

Economist Travis Fisher says, "The notion that wind and solar are becoming cheaper than sources like coal and natural gas is patently false."[69]

Shall we look at the capital export issue? After all, 76% of wind industry component manufacturing happens abroad. We can look at capital export issues, but it will bore all but a few of you. Know that capital export is twice costly: when it leaves the country and because it tends to stay outside of the country. Loss of gross, net, and government revenue ensue.

Let's just slog through more data on jobs. We will move on to regulatory issues soon. Please understand something critical about numbers. They lie. That is, it is remarkably easy to turn a figure into a statistic. Mark Twain said "lies, damn lies and statistics." What follows from both the government and this author are numbers transformed into story. They are used to make a point rather than tell the truth.

Jobs are one of the most important economic considerations for the energy industry. For example, during the 2008-09 recession, the oil and gas exploration and production industry created 1.7 million new jobs. This happens to be very close to the total number of new jobs created by the entire national economy during the same time period: 1.9 million.

What of the wind energy industry? It is interesting that the AWEA (who we met at the intro to this chapter) is the only source of jobs data for the U.S.[70] Their statement for total jobs creation in the wind industry is through 2010: 75,000 industry jobs.

Consulting, finance, logistics, transportation, engineering 45,000 60%
Manufacturing 15,000 20%
Installation, O&M 15,000 20%

The NAICS statistics for the nearest job classification, *turbine and turbine generator set units* (333611),[71] shows 26,218 jobs for the 2010 reporting period. The challenges to these figures are voluminous. They are well documented by Lisa Linowes.[72] Nevertheless, let's accept these figures, as we have much wood to chop.

Economists have used standard input-output analysis programs for at least 40 years to determine the positives and negatives of various economic activities. Numerous studies, using such economic analysis programs, performed in Spain, Italy, Denmark, England, etc., show for every job created in the renewable energy (RE) sector, about three jobs are destroyed in other sectors.

For every three green jobs created in the private sector, one job is created in government, and, as a general rule, for every job created in government about two jobs are destroyed in the private sector, largely due to added economic inefficiencies; no one should claim government is more efficient than the private sector.[73]

For every nine jobs created in the RE sector of the European economy, 27 jobs have been destroyed in private enterprise and three government jobs have been created, which results in a further loss of six jobs. Twelve new positions at a cost of 33 jobs. Such a deal.

But wait, the losses are still mounting, accounting friends. Wind energy jobs cost $329,000 each, according to The Manhattan Institute.[74] Let's do the math.

$12.18 billion in tax credits for 2013
37,000 jobs at risk according to AWEA
12,180,000,000 ÷ 37,000 = $329,189
$329,189 ÷ 10 years = $32,919/year

Factor these costs over a decade—if for no other reason than to amortize the tax cost just as you would the capital cost. Of course, no government official amortizes the cost of a tax credit, but let's humor ourselves. We get $32,919 tax cost for a job. This is spent each year for ten years. Tax revenues are forgone for this amount—that is what a tax credit is: a reduction in tax liability.

Now, the Congressional Budget Office (CBO), estimates that the oil and gas industry receives tax preferences of $2.5 billion each year. These are not tax credits but accelerated deductions, amortization deductions, and intangible drilling cost deductions (reduction in reservoir inventory). As such, they do not have the value of a tax credit. They are a *below the line item*, while tax credits are an *above the line item*. Nevertheless, let's use them as a comparison, unfair though this may be. The American Petroleum Institute (API) estimates that a loss of all of these deductions (a real legislative possibility) would destroy 40,000 jobs.

> Let's do the math:
> $2,500,000,000 ÷ 40,000 = $62,500
> $62,500 / 10 = $6,250/year over 10 years

As hard as it would be for the families of these laid off workers in either industry, from the viewpoint of DC, there is a significant cost differential: $32,919 vs. $6,250.

These figures are, of course, fantasy. One can arrive at any figure you wish. Susa Combs, the Texas State Comptroller, said in 2010 that each new wind energy job cost the state $1.6 million in lost revenue. Some extreme commentators arrive at even more egregious numbers, as much as $14 million per job.

Examining the 2009-2011 $9 billion Section 1603 Treasury Grant Program, the National Renewable Energy Laboratory (of the Department of Energy) figures 910 direct permanent jobs each year and 4,400 indirect permanent jobs over the assumed 20 to 30 year project life.[75] Total permanent jobs created over 30 years: 159,300.

> Let's do the math: $9,000,000,000 ÷ 159,300 = $56,497.

Each job costs $56,497. This is simply the cost of one of several tax credits available to the wind industry. The program used to calculate these figures, JEDI, is a knight errant, as it builds these figures from assumptions rather than real data. It also shows 40,000 to 60,000 temporary jobs in construction, etc.

We have attempted to show two ideas: 1. calculations can be manipulated, and 2. numbers are often imaginary. "Ah, what tangled webs we weave, When first we practice to deceive."

Robbie Burns was an excellent economist and legislator, eh? Of course, the Spanish estimate is quite different. The report from King Juan Carlos University shows a net reduction of 4.3 jobs for each new MW of wind energy.[76] Whether it is jobs lost or capital cost, the price tag for wind energy jobs is egregious.

The jobs story is deeper than that to which these forced cost comparison figures allude. Green jobs are often referred to in the alternative world. What is a green job? There is no actual definition from the NAICS. We could describe the arena in which someone might work: installation, record keeping, regulatory oversight, administration, and management. The beauty of a green energy solution—solar, wind, biomass—is that once it is built, there is typically less O&M (operations and maintenance). Compared to conventional energy sources, it is minimal. This is the beauty—and the job problem. A few ground site technicians, aka *wind walkers*, can do most of the maintenance. Three guys can cover an entire tax farm. The rest is paper pushing.

As we will see, market incentives for smart energy use at the household and business level are paying off nicely. Incentivizing reduced demand during heavy use periods very often works in reducing demand. The improved efficiencies that result far outweigh the alternative energy solutions in terms of capital cost and environmental cost. Reduced demand means less coal burned; it means less CO_2 in the air. Quite an elegant solution, really. Encourage end-users to reduce their demand by paying them to do so—or increasing the cost if they chose not to do so. It worked rather well in California for urban water use during the current drought.

Our intention is not to comment upon the politics of energy production. The author will leave that to others. We are discussing the myriad economic details. The devil is in the details.

Externalities

A full explanation, if not an accounting, of wind produced energy must include all *externalities*. This is a term coined by the alt-en industry to avoid certain *capital risk events* such as pricing mechanisms influenced by cost. A *tax credit* is a negative cost (a benefit) externality. Rare earth mining fatalities, avian deaths, and human health concerns are negative externalities, as per the wind industry. Current analyses do include these consequences for coal, nuclear, and gas. Oddly enough, they are excluded for wind. Sorry, we can't include these costs. Just move along. Nothing to see here.

If treated as the industry wishes, these externalities are excluded from a whole systems cost analysis. Their absence improves the bottom line. Ask the folks at Enron about exclusions. If you have a tax accounting background, you know to read the footnotes to any firm's P&L and Balance Sheet. What is hidden here are the externalities. The sins of financial management are buried in the piles at the base of the data source. These are *off balance sheet* externalities. The oil and gas firm that employs these does so as its own financial risk. Just ask the men at Linn Energy. Yet these issues are acceptable financial behavior for alt-en firms, particularly wind farms. Broken financial promises.

The economics of wind energy is less than ideal if it is measured on its own merits, without regulatory fiat, tax subsidy, or externalities. The advantages flow from its preferential treatment:

> its first in line status
> its enforced premium of more than two cents per kilowatt hour
> its accelerated write-offs
> its tax credits
> its regulatory bias

All of these in toto make wind energy profitable to the developer. The end user, the family, and the commercial or industrial consumer find a different result. Rates charged to these consumers continue to rise. Tax revenue is siphoned off to support each of these wind projects—which grow in size and number each year. Tax rates must also grow to support these avatars. A few wise energy purveyors, utility companies, and developers do quite well at this energy game. All other consumers and taxpayers must pay to stand at the end of the line.

What about the economics of incentives? Energy is central to the economies of every nation. Energy, like food and shelter, is an elemental aspect of any economic relationship between humans. There is a similar *basic market incentive* to reduce cost. This simple economic concept applies to virtually all financial decisions. First users of a high tech TV pay more while driving the public towards use. As more buyers approach the marketplace, more product is manufactured to meet the new demand. This drives down prices as costs plummet with greater production. More consumption leads to greater use at lower prices. Eventually a commoditization of pricing may ensue. Most internet successes are a result of this process: free information is made available. Advertisers tack on their wares. Sales ensue. The commodity is free information. The sale is in the add-on.

Certainly, the shareholders of all energy companies earn profits and share dividends when more use creates greater profit. Like it or not, these are the rules of the game of capitalism. Invested capital employing intelligent labor earns a fair return upon each. Higher wages, profits, and taxes ensue. The community is bettered.

If the wind doesn't blow, or blows intermittently, or blows too gently for too long, the results are disastrous for stakeholders of wind energy companies. These wind speed reductions are happening as you read. Why? El Nino. The wind energy industry is blaming the wind on its poor performance delivering alternative sourced energy.

Wind represented 4.4% of U.S. power supply in 2014. In Texas, it is 10%; in California, it is 7%. The U.S. is second only to China in use of wind power technology. Despite a surge in wind farm installations of 800% for 2014, power capacity has increased by 9% for the first six months of 2015 while delivered electricity has dropped by 6%. Wind turbines are running at 1/3 of their capacity.

A chart from Australia proves the point. The wind has dropped across the planet. Utility companies are suffering from the wind drought, as evidenced by this chart.[77]

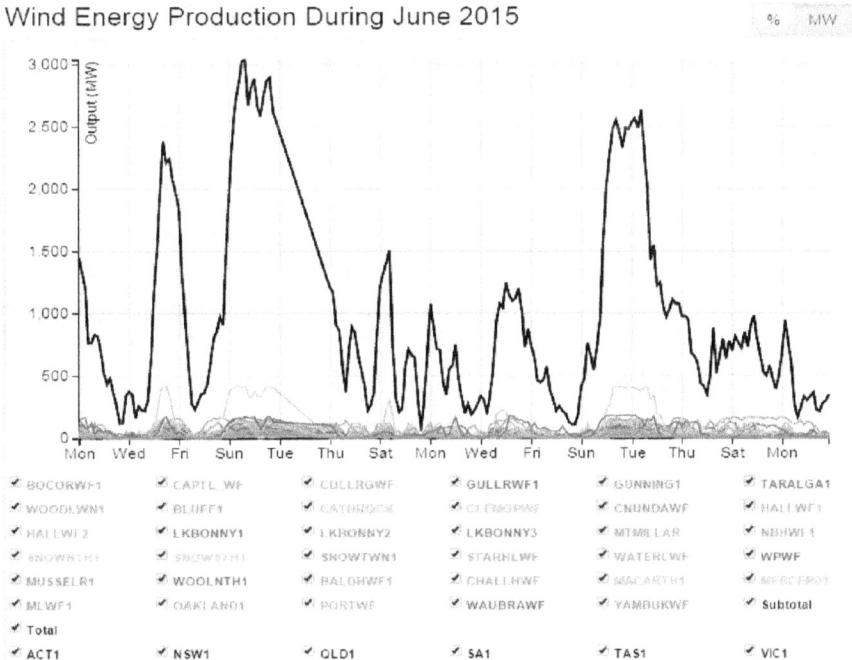

Wind Energy Production During June 2015

Capitalists get hurt and cry wolf, too. The wind bonds offered in the U.K. in 2011 were to pay investors 7.5% per annum for four years. There would be no trading market for the bonds, so holders had to wait until maturity for a chance at capital recovery. It seems that these bonds are now in default, paying no interest with no access to borrowed funds. Of course, these bonds were designed for the small investor with 500 English pounds to invest. Guess who got to hold the empty bag.

Wind Prospect said that like most renewable energy companies in the U.K., it had reviewed its options following the Tory attack on onshore wind. In order to minimize the impact of government announcements for its ReBond holders, the business is proposing to separate its services business and development assets.

> *"The existing U.K. and overseas development assets will then be ring fenced so that the proceeds from these are directly and contractually available to pay interest and repay capital going forward."*

To achieve this, *Wind Prospect* has asked its bondholders to agree to a three-month moratorium on payments of interest and capital while the details are confirmed and a productive consultation can take place.[78]

Here in America, shares of green energy companies have been the darlings of the stock market—until 2015. Now they are the dogs. Shares of the spinoff SunEdison have collapsed as they are unable to meet their debt obligations and pay their expected dividends. As a *yieldco*, this type of firm is specifically designed to earn sufficient income to pay decent dividends to shareholders. They have recently laid off 15% of their workforce, over 1,000 workers. They have had a margin call on their debt as well.[79]

In the U.K., British wind companies are protesting loudly against the removal of all subsidies for wind energy by 2016.[80] The Renewal Obligation (RO) will expire 3/31/16. Coupled with the elimination of offshore wind energy support, this has decimated the industry. Why would this be, if it is a viable business model designed for success? Why must it have government support or die?

In the past, as energy is provided and used more efficiently, costs decline and usage increases. The drop in cost is advantageous to all, often in reverse proportion to their wealth. Those in the lower deciles of financial society find they can afford energy as its price declines. They can use more for other things.

At the human level, the poorest receive the greatest advantage from any price reduction. Why? Energy is a greater share of their total family expense. *A small drop in energy cost is doubly magnified for the poor* as it adds more to their bottom line. A $300 monthly electricity bill represents 14% of the income for a family earning $25,000 annually. It is less than 2% for the family earning $250,000 annually. A reduction of 10% ($30) means more to spend elsewhere on food and fuel. The 1% reduction in outgo for the wealthy family may be of statistical interest only.

It also aids the poorest by its second salient—increased usage. Cheaper energy allows for more application to other uses. Education, food, shelter, health, and legal rights tend to increase as energy costs decline. This is an observed phenomenon, fairly universal in application. **The poorest receive the greatest benefits from energy cost reduction**. Every family seeks to provide for its survival. It does so selfishly. Adam Smith's *invisible hand* grasps Darwin's hairy fist in all societies. Each act of selfish interest spurs survival of the species. Man "intends only his own gain and is led by an invisible hand to promote an end which was not his invention."[81]

This is both a scientific fact, and an observable phenomenon. It works in a sufficiently large number of cases to make it noteworthy. It can however fail if other costs impinge upon these two important degrees of improvement: price and usage. Clan or governmental imposition, cultural redaction or political depredation, can overwhelm the impact of lower energy costs on any nation's poor families. This is as true as the obverse. Governmental subsidies or price support for a universal item such as bread become both addictive and intractable.

Taxes and regulations (and religious or social pressures) can destroy energy cost savings advantages. Harry Alford, the president of the National Black Chamber of Commerce, told the Senate Judiciary Committee's Oversight Subcommittee:[82]

> *"The proposed Clean Power Plan would impose severe and disproportionate economic burdens on poor families, especially minorities. The EPA's regressive energy tax threatens to push minorities and low-income Americans even further into poverty."*

Alford cited a study that his group had commissioned. It found that by 2035, the Clean Power Plant would have boosted energy costs for blacks by 16% and Hispanics by 19%.

As with bread, when the price of a commodity such as energy is artificially suppressed, it becomes an expectation. It no longer has social value. The opportunistic value of falling prices resulting in greater use is replaced with the stagnant expectation of a free ride. The urge to achieve more is stifled by the absence of value.

"You owe it to me" replaces "I have earned this." Harry Alford speaks eloquently against this regressive tax burden and societal racism.

John Droz feels that any energy source should be available provided it offers "a net societal benefit."[83] In a society where cost is dictated, changes in the actual produced product price are immaterial. Efficient use of capital becomes redundant. Profits are subservient to preferences. Tax preferences allow continuity of a political process. This process is parasitic. Each infests the other, sapping life from both. It is addictive and extractive. Capital feeds politics. Politics feeds capital. This removal of capital from a society's cash flow via taxation and regulation destroys the wealth of such a society. Even if some capital is returned to end-users in the form of subsidies, much is wasted or simply stolen. Never does it all return to society's balance sheet. Much is lost forever. Net societal effect? Never saw it. Don't know it. Can't speak to it. What is it, anyway?

This is true for any commodity, any price support mechanism, any intervention into a free hold marketplace. Sand in the cogs of a machine, whether from oil interests, solar interests, hydroelectric interests, shit interests, or wind interests, always result in economic inefficiencies. This sand always results in enhanced poverty for the least able to respond. Rent seekers become close allies of politicians. They write the rules by which they are governed. They propagate their tool of exploitation. They design their own incentives for success. The game becomes rigged. Fear and desire becomes tools in a spiral descent into disaster for the poor, for the unrepresented, for the disadvantaged.

The center cannot hold.

Let's look at the same story from a capitalist perspective. Recall that reduced pricing tends to aid the poor via free cash flow spent upon items of interest rather than of need: education, health care, and legal rights. In a less aggrandized economy, the opposite of reduced price and increase usage results when a commodity is provided at an artificially high price tag. When this price is artificially high, it reduces opportunities for certain members of society—the poor, women, minorities, and youth—by reducing use even as price increases. This casts incalculable pressure on workers, whose wages may not be responsive to changes in energy price enforced by regulatory dictat. Inflation or reduced use follows. Poverty flows from such inflated expense.

There are winners certainly. Tax credits flow to the top 10% of the population, those in the highest tax brackets. They naturally consume credits just as hogs consumer fodder. These credits increase their net income. The Tax Code is designed to advantage the wealthiest. Its subterfuges and machinations are best advantaged by tax accountants and lawyers. Complexity breeds self-awareness.

10% of the taxpaying population of California consumes 95% of all tax credits. Energy tax credits are one of the most efficient drivers of income inequality. They force utility costs to rise, take a larger share of the poor's take home pay, and result in lower usage at higher cost for the disadvantaged. This is fully one third of the resident population of California—those living below the poverty line.

In Europe, the price of electricity has tripled over the past decade. While some argue that wind power will soon be cheaper than natural gas power, they have been making this argument since the 1980s.[84] In Germany, wind energy price is $80/MWh, while the spot price for natural gas power is just $33. These plants are closing. The ones that remain must run at half speed to back up the turbines when the wind doesn't blow. In Spain, the gas plants run less than 60% of the time. As much as 2/3s of these plants may close in the next decade. The cost for electricity is headed skyward.

The world produced 635 TWh of wind electricity in 2013 at a subsidy cost of $28 billion. This is $76/ton of *assumed avoided* CO_2 emissions. The carbon price in London is $5/ton. It is cheaper to buy an offset than to produce from this Faulty Tower. If we do everything the IPCC expects of us, the world will spend $2.5 trillion to increase alt-en sourcing by seven times. The result? According to the IPCC mandarins, the net CO_2 benefit will be a reduction in global temperature of less than three one hundredths of a degree Fahrenheit: .03 degree.[85] Never has so much been spent to do so little.

Poverty flows to the least advantaged via inflated expenses for basic necessities like electricity. Inflated expense is precisely the result of preferential commodity pricing. The marketplace is distorted by ideological principles in search of authoritarian virtue. Their authority comes from the future. Environmentalists force feed us their liturgy of catastrophe. All will be lost soon by our wholesale destruction of the planet, of everything we have ever touched. We are unclean. We must purge ourselves of these sinful acts. We must renounce once and forever the Hydrocarbon God. Only our own self-sacrifice will make Gaia whole again.

What of the world's poorest of the poor? They suffer the greatest from economics imposed from on high. The world of Aid Donors is a case in point. Bono, Melinda Gates, and many others have suggested wiser allocation and use of the billions on annual aid to the poor. They urge small projects, tightly focused objectives, and deeper involvement of local people.

Big donors still abound. Like dinosaurs in the Paleozoic Forest they dominate with their appetites. What do they desire? Big Projects. Big Names. Big Ideas. Millennium Goals. Their fashionable ranks of plesiosaurs, brontosauri, and triceratops are now in need of Climate Aid. It is soo coool to give solar panels and wind turbines to the Wretched on the Earth.

Fanon would be ashamed of the exploitation. The Algerian Marxist knew what the people of the world want: clean water, inexpensive and nutritious food with legal rights for all. They get unwanted solar panels and used turbine skeletons. $100 billion is promised by the rich in climate aid. One half of one percent of that amount—just $570 million a year—would reduce malarial deaths by 300,000 annually. That is a 50% reduction. Such a shame.

When asked by the UN each year, the poor rank climate action at the bottom of 16 priorities. Eight million participate in this poll. They prefer food, water, and nutrition to solar, wind, and biomass. Three billion people still cook in the room where they live, eat, and sleep. Indoor pollution is their biggest challenge. Access to natural gas powered electricity would very quickly eliminate their health, safety, and food concerns. According to the 2014 Center for Global Development, 60 million people could gain access to electricity if the Overseas Private Investment Corporation, the U.S. federal financial institution, "were allowed to invest in natural gas projects, not just renewables."[86]

We have ignored this environmental cost in past cost accounting exercises, certainly. Yet how do you determine such a cost? In reference to wind energy, to any form of alternative, renewable energy, this cost is universally defined in terms of future results.

The a priori assumption? Our health, air, water, natural resources, and other species, each and all shall be degraded if we take no action. The threat of inaction is primordial. It is the new Original Sin. These imminent results are known. They are unquestionable. If you refute this, you are a denier, as despicable as a Holocaust denier, a shallow skinhead with no underlying hippocampus. Your reptilian remnant is the stalking horse of your guilt. RICO threats are now being suggested for such irredentist thinking. These would enforce a fine of three times the penalty on those who deny the climate change hypothesis. Let's do Mao. Let's mao-mao the environment.

Since these future climate results are defined as true, then the costs associated with these changes must be learned. They must be ascertained. While a recent meta study has found more than 2,400 factual errors in the dozens of climate models being tortured,[87] facts are subservient to fear. Entrails are consulted. Models are formulated. We must try to determine the economic extent of this disaster to quantify the financial cost of inconsiderate inaction. We must urge the ignorant masses onward into the abyss.

How can this future result be known? What are the mechanics of such knowledge? *Computer models* are the study of the future every climate scientist worth his salt employs. These models are the *sina qua non* of climatology. They attempt to imagine the world of the future resulting from the continued spewing of hydrocarbons into the planet's air, water, and dirt.

Initially, they imagined a world gone cold with industrial ashes and gases. In the 1970s, the Earth was about to freeze over. Not, of course, before we had exploited the last remaining resources and plunged humanity into a death spiral of starvation and cannibalism, ala Paul Erlich. As time passed, visionaries portrayed for us a slightly different reality. The models were changed. They were updated. Now they show an increasingly hotter world resulting from the evil gas, CO_2. The figure recorded from atop today's Mt. Olympus (Mona Kea in Hawaii) shows a slow steady increase from 350 ppm of CO_2 in the atmosphere through 400 ppm at today's count. Deeper Blue computes the results for the planet as we destroy her, gash her, rend her. More power and more data overrun by more computing power give us the road kill of today's Flattened Earth. Avatar rules.

This *oology*, today's evolved version of the study of animal entrails, is force fed to the media. Computer models foretell the future with perfect precision. Such future accuracy is amazing to behold. It must be true, because it is so bold. The degradation of Earth's entire environment is expected if we do nothing. We must renew our energy sources and consumption habits. We must change. It is a self-fulfilling prophecy. 97% say it's true. Or at least that is what you are told.

Any cost associated with this change is small in relationship to the alternative, the wholesale destruction of the biosphere of an entire planet by one excessively greedy species. Much has been written about the expected environmental cost of, and the savings resulting from, radical changes to the social experiment of Man. All of these changes are warranted, as less costly than extermination. Many others have suggested such an extermination as the last best hope for Planet Earth.

Thus, the massive increase in the cost for power in much of Europe is a small price to pay. The wanton destruction of the lives of Chinese miners and their families through extraction of rare earth minerals is worthy of the cause. Birds will survive our onslaught, if only barely. The alternative is their entire destruction as a genus.

The Stern model and the Nordhaus model, as evidenced in a recent editorial in *The Economist*,[88] give the most heinous foretelling of our ignorance. Using mathematical formulae that discount future benefits at a negative rate (a mathematical impossibility), they arrive at revelatory conclusions. The model assumes that economic growth rates will be destroyed permanently rather than affected in the interim. Thus all future damage is currently understated by today's low discount rates, even those as low as 1%. The effects of climate change on the global economy are underreported. These models need to deny *in toto* any discount rate. They must assume an exogenous (natural) negative growth rate. Negative discount rates are the only solution. Only with these mathematical aberrations can the newest models accurately foretell the immanent future destruction. Never mind that a negative discount rate is an impossibility.

The model must be correct. If it is not, then change the figures to make it correct. It is perfect; we simply have to make it more so. Some are more equal than others. Ptolemy was an optimist. Cambodia, Rwanda, China, and The Holocaust are but intimations of what is yet to come. *The Horror. The Horror.* Were Kurtz's prophetic words. Simply a preamble?

In today's real world of economics, the cost of energy can be positively or negatively impacted by an externality. Regulatory forces can ensure scarcity and thus force prices upward. Market forces can do the opposite, ensuring greater supply and reducing price. Predicting any of this is fools' play. These childish games have been played by knowledgeable researchers since economic research began. Their predictive skills are universal: They are universally wrong. Trends are missed, or misread. Estimates are significantly higher or lower than reality provides just a few years downstream. These guess mongers are always consulted, regardless of their complete lack of success. Oology is alive and well in Economics Land.

In the real world of your life, your family, and your environment, if you choose to be concerned about an increase in CO_2, you can follow the recommendations at the end of this book. They are easy and doable by all. You can reduce your carbon footprint if you so choose. If enough do so, the impact may be noticeable. You will be acting as a good steward of your personal natural resources.

Nevertheless, the supposed effects of an increase in CO2 have yet to be observed. Weather is not climate. More people have residences on beaches in warm climates, thus the costs of hurricanes has risen. The frequency of hurricanes has declined, as has their intensity. Winters are colder, summers are drier in some parts of the country (and the world). The Arctic ice has sometimes more, sometimes less ice of greater or lesser thickness. Greenland has more ice than ever. The Northwest Passage was open every year during the first quarter of the 20th Century. There are more polar bears today than have ever been recorded.

The forested acreage of the Northern Hemisphere is greater than it has been at any previous human era. The number of tress on the planet is now estimated at three trillion. This is eight times more than was estimated just two months ago. This is an error of nearly an order of magnitude in size. This is no rounding error. Trees absorb CO2, just as we exhale it.

Solar cycles have a closer, more immediate, and more serious impact on hemispheric climate change than any other source. No model has yet predicted weather, much less the climate, further out than 3 to 4 days. None has aptly predicted climate stress, much less change, ever. Barbie is a model, she is not real. Climate models are not real. Economic models are not real.

Models are meant to be the beginning of a conversation, not the end of all discussion. The simple truths are...

> The future will not look like today.
> Innovations cannot be predicted.
> Their impact is unknowable.

> Researchers' biases impact studies.

For the purposes of this book, we shall ignore these ruminations of the future. We cannot know it, we cannot foresee what our children will create nor can we know their individual and cumulative impacts. We can take these fulminations as yet another **broken promise**, a promise for a future unfulfilled. These promises are the lowest form of lies. They distort the truth. They ignore their own weaknesses. They suggest an ideology if not a religious belief system. They are fool's gold.

Let's take an example of simplistic economics applied to our subject, wind. *Wind is free.* So goes the mantra for the industry. At one level this is quite true. It is also true about every form of energy extraction: hydro, carbon, nuclear, bio, solar, or shit. The source is free. You do not have to buy wind, water, crude oil, methane, fission, sunshine, or shit. Each is free.

At each intervening level, this becomes clearly untrue. The extraction is costly. The processing is costly. The transmission is costly. The end use is costly. Construction, extraction, processing, storage, installation, O&M, transmission, and distribution all have costs. These costs apply across the energy spectrum. Hydro, carbon, nuclear, shit, and solar must incur these expenses to get to your home from the free source. These costs are inherent in the structure of delivered energy. We extract it from a passive source, revolutionize it from kinetic energy to applied energy, and offer it to all users at a (hopefully) fair price.

Sorry to burst your balloon. No freebies in the wind.

By the way, shit is a source of fuel for tens of millions across the globe. It produces horrific smoke in small enclosed spaces like primitive huts and encampments. It kills far quicker than any tobacco product. It maims women and children first. It is the first and worst form of energy. Freeing humanity from shit-based energy is the highest goal of any energy program, or it should be.

From a wind energy economics point of view, the price advantage of wind lies entirely within the Tax Code. The final evidence for this is exemplified when wind producers pay users to take their energy—they sell at a negative price. Why? They only receive the federally mandated production tax credit if the energy is actually sold. As they get 2.2¢/kilowatt hour from the feds, they can pay you (actually someone else) 1¢ and still make 1.2¢.

Let's do the math: 1.2¢ x 600 MW = $7,200 x 365 days = $2,628,000.

This is a very simplistic description. It demonstrates the point. You can sell at a loss and make money, very good money at the deal. Nice work if you can get it. Ask your local political representative. The game is rigged. You cannot lose. Well, you the consumer do lose, but today's savvy wind energy executive cannot lose. He has become the house.

Absent this pricing advantage, wind would price itself out of all but the smallest and most remote markets. As with solar, there are excellent places in which wind rises to its full sail. Distributed energy in pioneer homes and remote villages is better supplied from locally sourced wind and/or solar. The most dramatic applications for wind energy today are at the small end of the technology scale. As we shall see when we look at HWAT alternatives, the home or backyard based wind tower can be quite well positioned to your advantage.

If you truly want to go *off grid*, today's small scale wind towers, generators, and blades can make all the sense in the world. They are shorter, so they do not interfere with bird migratory paths. They are smaller and slower, so they do not act as bird blenders. They are entirely free of the grid, so you neither sell nor use utility company's fictitious bitcoins. The energy they supply comes at a cost—as do all energy sources—and this cost is quickly recovered through the elimination of utility surcharges (those odd little items on your electric bill each month). You become the house. As long as you and your neighbor don't mind the sound, as long as you have an alternative power source to answer the challenges of intermittency and fluctuation, you are home free.

Where is this ideal haven? At sea, of course. Many of today's global sailors generate all of their power supply from a combination of wind and solar. They have plenty of both, as most sailors tend to travel equatorial rhumb lines. They use DC electricity, so their simple batteries can hold the power supplied as an energy bank. Blenders, sewing machines, microwaves, TVs, computers, and radio all work quite well on DC run through an inverter. Rare is the anchorage without an afternoon breeze to top up your battery bank each evening—just in time for margaritas, music, and mahi-mahi!

Remote Alaska and Africa are equally advantaged by wind and solar. The remoteness adds to the advantage, actually. Ease of mechanical application reduces O&M to a minimum. Sun and wind are in abundant supply, albeit with the usual minima of intermittency and fluctuation. Poverty is quickly and easily abated with these cheap energy sources. Add some simple cooking technology as we discuss later and you have the ingredients for prosperity on the Sahel, the savannah, or the Sahara. Alaska offers remoteness, extremes of climate, a plentiful food supply (as long as you do not become the supply), and a certain eccentricity of the human soul to add spice.

Your point is simple. *From each power source, demand its inherent strength; to each energy need, demand a compelling power source.*

We have some good economic news on the wind story to close out this chapter. The industry is getting hammered by the markets. European governments are withdrawing or withholding legislative and regulatory tariffs. People are tired of paying excess electricity prices for the imaginations of a power clique.

> The industry is getting hammered.
> Tariffs are being reviewed.
> People are sick of paying for the dreams of others.

CHAPTER SIX
BUZZARDS

https://www.youtube.com/watch?v=C6OGp1tGCX8

Unlikely to win any beauty contest, buzzards are the trash collectors of the planetary plains and fields. The New World species, Catharditae, is not endangered, as it proliferates over the entire North and South American landmass. We had eight on our fence posts each winter morning splaying their wings to dry in the early sun. A few hundred would roost in the oaks each evening, maintaining a certain pecking order of which we were unaware. Huge in size but actually quite light on the arm, they weigh less than five pounds. Aerodynamics keeps them aloft as they soar on thermals with no discernible movement other than a feather adjustment.

Like their Eurasian counterparts, they are carrion eaters. They keep the area clean of road kill. They watch for ground based hunters and land, waiting their turn. They are featherless on their heads to allow stretching into body cavities. Thus, they are ugly.

I once drove across the Sahara, stopping in towns and villages for supplies and meeting the many tribal inhabitants of the Sahel. While in Burkina Faso, we crossed the Niger River on our way to Timbuktu. Ouagadougou was our destination. The market was a corrugated roof held up with palm stumps. The stalls were run by women sitting in the sand next to their wares, as is often the case in Africa. We bought what we needed for the next leg across the desert ergs.

A recently killed goat wrapped in a rag nestled beneath my right arm dripping blood. Following behind me quite closely were two elderly midgets in filthy tuxedos. They snapped at the meat with long extended beaks, fearless of my desperate moves away from them. These desert vultures were doing what they do best. Cleaning up after messy humanity.

The North American cousins bear no relation to vultures. They may share very distant branches from Asian bridge crossovers or much further back in time to Gondwana. No one seems to know.

Regulations

The ignorance of a plain man who knows where it hurts is a safer guide than any rigorous direction of a specialized character. — *Winston Churchill*

A *must take* resource is one the stands at the front of the line when it comes to power generation. Virtually all alternative fuel power sources are must take. This mandate can be federal, state, or both. Incentives are put in place to encourage compliance with these regulatory dictates. As an example, the price of wind power is juiced by 2.2¢. RPS are the political tools to ensure these must takes are enforced. The Chicago mafia stands guard over the futility of our future. They will extract their pound of flesh for the benefit of themselves. The next generation will certainly carry the weight of these unpaid bills of tax lading.

The tyranny of technology assumes that practitioners are the only ones available to fix the shortage of expertise. Freedom of choice, freedom of application, the very freedom to fail is far less important than learned skills, preferably from an ivy bound enclosure.

The process of failure leads to results, either a complete dead end or a simple, elegant solution. Failure is the modus operandi of the gene pool. Because it is assumed, there is a surplus of achievers. Evolution works this way. Human thought works this way. Government, heavy handed capital, and ideological intervention work against this simple principle: failure breeds success.

For the disparate crowd, only their words in the form of regulations will achieve. The regulatory process tends to choose a particular technology and stick with it. Other choices outside the square are less valuable. They are often denied important opportunities to prove themselves by failing. Certainly, the capitalist system of intense competition is heartless. There is nothing easy about cutthroat business activity. You struggle to make it or you fail. Bankruptcy is the typical outcome for the majority of enterprises. Yet, the competitive approach—the jungle mentality—is precisely why progressive society moves forward. Making difficult choices hones the blade, strengthens the spirit, and increases the odds of ultimate success.

Sticking with the model works for government types. There is no square outside. There is only the regulatory box. "Fill in the box completely, leaving no edge unmarked to complete each answer. You have two hours to finish the exam."

Central planning in whatever form is just that. It urges caution. The *precautionary principle* is the perfect regulatory ideal. "If an action or policy has a suspected risk of causing harm to the public or to the environment, in the absence scientific consensus that the action or policy is harmful, the burden of proof that it is *not* harmful falls on those taking an action."[89] Take no chances. Green jobs programs consumed $20 billion in federal tax revenues in 2009. All of these dollars will be borrowed from the future and paid thrice over by our children. How many jobs do we have? 75,000 in the wind energy industry, according to its own spokesperson (lobbyist). A total of $63 billion was spent directly on green energy works in the same 2009 stimulus bill. According to the White House, the figure was $80 billion+.[90]

When will we stop the madness? When will the broken promises be replaced by heartfelt attempts? Where are the heartrepreneurs? Those willing to do their best in the interests of society's best, or the planet's best? Who will step forward to break the cycle of idiot dependency upon foolish old men and their ideologies?

Spain

Spain has been impacted significantly by the wind power events of the previous decade. Jobs, subsidies, towers, and taxes: each has grown to spectacular heights on the Andalusian plain. She has taken the tablet. She now has as much as 25% of energy generation from renewables. The cost? That would be a $41 billion short fall between cost to produce energy and allowable charges for the same energy. The difference is picked up in the form of subsidies. Better known as taxes or, in Europe's case of sharing the poverty, infusions of capital from Northern Nations.

Jobs impact was the focus of a 2009 study from the Spanish government.[91] The cost per wind industry job exceeded expectations: €250,000. The number of permanent jobs declines drastically once O&M, operations, and maintenance take over the job site. A dozen workers can service dozens of turbines. The costs were completely out of proportion to the benefits.

Germany

Recent German Green party internal e-mails leaked to *Der Spiegel* magazine show government subsidies in excess of €5 billion Euros/year have not led to a reduction of a single gram of CO_2 emitted on the continent of Europe.[92]

The retail cost of electricity in Germany has risen by 17% in four years. The feed-in tariffs force the use of renewable sourced energy, without incurring the direct cost. Again, subsidies replaced marketplace pricing mechanisms. In Britain, a large steel mill recently closed, forcing 2,000+ workers out on the streets. Why? The firm was forced to pay a premium on its electricity usage because it was sourced from wind. Jobs? Gone.

The European Commission abandoned its country by country targets for GGE emissions reductions. These were fig-leaved with a 40% reduction across the continent by 2030. Everyone remains committed to the goal yet no one is responsible for its achievement. The *Tragedy of the Commons* once again. National targets are being eliminated because of the massive economic pain suffered from renewable energy dictat.

In Germany, the cost for wind energy is widely accepted as three times that of conventional energy. Household electricity bills have doubled over the past decade and may triple again by 2030. What is the price for the assumed value of carbon reductions? €1,050/ton of CO_2. When extended to such alternative fuels as biodiesel, the production cost is four times the price in CO_2 that conventional diesel production represents.[93]

The results of this regulatory morass are clear. Germany's largest solar manufacturer, Q-Cells went bankrupt in 2012. Solar-World hangs on the cusp of destruction. Prokon, the German wind park developer, went bankrupt in early 2014. Vestas has had to reduce its size through restructuring from cost cuts and massive layoffs. The European wind energy consortium has had the worst performance of the stock indices. Mr. Market is never wrong. German utility firms have lost money for the first time in 60 years. That is very difficult to do when the table has been set for you for three generations. It takes a true muck up to starve the elders in the household.

The reasons offered by these industrials are the lack of serious, binding renewable targets for the continent through 2030. They just aren't getting enough tax support folks. If they had these compulsory targets, they could produce wind turbines en masse to meet the dictated demand, thereby achieving, finally, economies of scale. The utilities could make a profitable transition if only the end user had to pay a fair price (they will set the price, thank you). Meanwhile, GE and Siemens have overtaken the turbine market with large scale projects and massive turbines. Build 'em bigger, boys.

Did you note the complete lack of concern for the end-user? The industrial, commercial, and retail consumer are kept in the dark. "Ya pays yer money and ya takes yer chances, fellas." Not a word is mentioned about the increasing CO_2 emissions from the Eurozone. They burn more coal and wood so they don't have to burn uranium atoms or methane molecules.

Bats and Wind

In North Carolina, John Droz has taken the data from an economic study on bats and applied it to the new turbines planned for the coastal counties of Carteret, Pasquotank, and Perquimans. The study is explored in detail in the subsequent section on bats. He takes from their Excel spreadsheet the median of the estimated cost of bat losses annually.

This is the only known economic work on bat losses from wind turbines. It was done by four PhDs: Drs. Boyles, Cryan, McCracken, and Kunz. It was printed in *Science*, 4/1/11. This is a peer-reviewed publication. Their back story is complete and well documented.

Their conclusions are spread out across a county map of the U.S., all 3,141 counties or equivalents. Droz applied the economic losses to the three counties soon to be impacted by wind turbines installation. The builder indicates a net pre-construction benefit of $1 million to $1.5 million for each county in the form of tax benefits to the counties and land lease payments to the landowners. John arrives at a figure of $20 million in annual economic loss, primarily from bat slaughter. He bases this upon the research from the biologists writing in *Science*.[94] He has a name for these activities of the wind turbine industry: **Greenwashing**.

A major turbine manufacturer calls a catastrophic failure like a blade falling off *structural liberation* (http://stopthesethings.com/2015/12/22/happenstance-or-enemy-action-giant-wind-turbines-collapsing-with-alarming-regularity). The same industry calls for the complete closure of all coal plants. Let's discuss this issue.

Cycling

Cycling is the process of maintaining proper pressure (voltage) in the grid as wind power comes online and as it slackens. It is mandated by *Renewable Portfolio Standards* (RPS), which have been adopted in 29 states. Let's examine these in action in Colorado.

RPS is the state requirement that alternative energy power sources be regarded as a *must take resource*. Conventional power sources—coal, natural gas, nuclear—must adjust their power output in accommodation of alternatives. The NREL[95] lists these potential benefits and costs of RPS:

Benefit	Cost
Fuel diversity	Higher electricity prices
Economic development	Higher distributive costs
Power price stability	Higher generation costs
Emissions reductions	Higher integration costs
Water savings	Transmissions upgrades
Grid security	

Their web site, called DSIRE, or Database of State Incentives for Renewables and Efficiency, is replete with reasons for alternative energy development. Facts are deployed in support of the transition to a post carbon world. Not to put too fine a point on it, but there are many citizens for whom this world may be less than desirable. Most of them are poor, illiterate, and living in shit smoke-filled rooms dying slowly of noxious fumes.

The measurable, real world result of RPS is called cycling, the sudden increase or decrease in power output. These changes to power output from coal and natural gas facilities increases their inefficiencies, resulting in the release of more, not less, SO_2, NO_x and CO_2. Greater cycling results from varying amounts of wind, both diurnally and as a result of weather changes, both stormy and sunny days. The wind blows unevenly across the land's surface undulations. Gusts and williwaws, calms and gales result. Each of these directly impacts the delivered energy from every wind turbine in the way.

The DOE weighs in on the cycling issue each year in its annual *Wind Technologies Market Report*. This is an excellent overview of the economics of wind energy, at least within the rigors set in place by the DOE. The 2012 report indicates:

Large *balancing areas* can ease wind power integration.
Coordination between smaller areas can ease costs.
Wind forecasting can reduce costs.
Power *scheduling and dispatching* flexibility can ease costs.
Conventional cycling contributes to cost challenges.
Balancing reserves can be 15% of nameplate capacity.

These integration costs can add \$1.20 to \$5.40/MWh to electricity cost. This wide variation in cost is indicative of the fledging nature of the wind energy integration marketplace. For many, this is an exciting time in wind power. The dynamics of grid integration across many sources, demands, and transmission facilities in nearly real time is the cutting edge of the industry. The same can be said for the whole systems integration attempts by oil and gas field developers. They have to process flow rate from each well, type of hydrocarbon, pressure, pipeline availability, storage capacity, and demand metrics, also in real time. The folks in whole systems management, such as Oracle, are having a field day in this technology universe.

Today unfortunately, the direct result of the use of more wind power is the release of more pollutants. Whether CO_2 is a pollutant or not is subject to political interpretation. For purposes of this chapter, we accept it as such. As most conventional power plants are sited near population centers, the urban results of carbon dioxide production can be hazardous, particularly to youth, the elderly, and the infirm, or so goes the literature.

Coal plants are very difficult to cycle. They are designed to run at stable flow rates, to generate reliable, continuous power generation. Natural gas plants and combined cycle plants can each do so far more readily and efficiently. This may be viewed by many as a further reason to accelerate coal plant decommissioning. Please remember the need for backup power to adjust for the intermittency of wind power generation.

As more wind generation power comes on line, the need for more backup increases. The backup should be designed to be natural gas and combined power generation sources. If these are built at a similar pace to wind power generation plants, and at the decommissioning pace for coal plants, then a progressive rendering of power sourcing can be achieved. If this wise transition is ignored, then two results will occur: more pollution and more power interruptions and brownouts.

Let's ask the following questions:

What is the environmental impact of these new wind farms?
What are the estimated climate impacts?
What are the effects upon avian populations and human health?

CHAPTER SEVEN
VULTURES

Many thanks to the brilliant work from National Geographic. Caracaras are a variation on carrion eating vultures. This video just begs for a story line and NatGeo delivers.

https://www.youtube.com/embed/Y7qcNiJTfVU

I stand upon the shoulders of giants, reaching for the stars. — Tycho Brahe

La plus ca change, le plus c'est la meme chose. — French proverb

Climate Change

The most important reason stated for the development of wind energy is *climate change*. While the author's views are clear, they are irrelevant to this discussion. We shall simply accept the fact that this hypothesis drives the ongoing pursuit of wind energy. By accepting it, we also accept all of the environmental issues regarding wind turbines. If you do not, or if you question these assumptions, then you question the very basis for the wind energy industry. You also subject yourself to immense ridicule, but that is your heroic decision to make.

The DOE's 2009 report, *20% Wind Energy by 2030 Technical Report* seeks to validate its title's claim. It demonstrates a resulting 25% drop in CO_2 emissions. It substantiates this with the observation of a 28M ton reduction in CO_2 emissions during 2007 from wind energy production.[96] Compare this to the 500M ton reduction in CO_2 from the substitution of natural gas for coal during the 2007-2012 time period. The parameterization of data in this report makes it a comic skit in one dimension. It simply illustrates the results of what would happen if 20% of our power needs came from wind power. It changes no other source, cost, subsidy or transmission function, leaving them all at 2006 levels.[97] The illustration leaves much to be desired and far too much to the imagination.

Let us try to accept these suppositions as factual. We then must try to substantiate the use of wind energy environmentally. What are the environmental costs associated with wind energy production? They fall into several categories:

Externalities of produced turbine materials
Constructions costs
O&M costs
Transmission costs
Wildlife costs

We have already seen that the physics of wind energy is less than perfect. You get little electricity from a free but difficult kinetic power source. Wind requires large amounts of invested capital and substantial, ongoing, decadal tax leverage. It is intermittent. It must have substantial backup. It interferes with other energy production sources by scrambling the grid, in sections or completely. The more wind farms are built, the more conventional power plants have to be built to provide cover. This leads to more CO_2 emissions rather than less.

The environmental impact of wind farms speaks volumes. *Neodymium* is an integral element in wind turbine generators. It is mined in California (was) and China. Mark Smith, the CEO of Molycorp Minerals (a California rare earths mining company), said in 2009, "We've coined the term *the green elements* ... because so many are absolutely indispensable."

http://www.youtube.com/watch?v=3NIVIOaQf2U

Green technologies—hybrid electric vehicles, wind power generation, permanent magnet generators, and compact fluorescent light bulbs—will not work without them.[98] Just one example: The two-ton permanent magnet on top of a large wind turbine generator is composed of 560 lbs. of neodymium.[99]

Would you like to visit Baotou province on the Chinese border with Mongolia? I doubt it. This is the center of operations for rare earth mining in China, which controls 90% of the planet's rare earth minerals resources.

http://www.youtube.com/watch?v=JL4fIuj004o

The 80,000 tons of rare earth produced in China yields 19 lbs. of fluorine; 29 lbs. of dust; more than 400,000 cubic feet of hydrofluoric acid, sulfur dioxide, and sulfuric acid; and tons of radioactive waste.

http://www.youtube.com/watch?v=yO94WHkqHg4

Ten million gallons of untreated waste water is then released into the fetid tailings lake of the province. This waste water either evaporates or leeches into the province's streams and rivers. These flow directly into the Yellow River, one of the most important riverine systems on the Asian continent, providing consumable water to 150 million citizens. These figures are taken directly from a 2009 report of the Chinese government.[100]

Safe *tailings* storage is non-existent.[101] The mines are worked with ancient open railway carts, with no semblance of environmental controls. Read the footnote's reference to the 2011 *Daily Mail* article if you dare. It is gruesome. The mining, the extraction process, the waste treatment all are ably described, as are the suffering and deaths of local villagers and their children. Cancer victims are in their thirties. Black lung disease kills thousands each year. It is the most common disease in Baotou province.

Where, oh where, are Rachel Carson's minions? Have you no shame? NGOs turn their heads away, fearful of the consequences of acknowledging the facts. Death, disease, malnutrition, and human suffering are the clear and present danger. Abandon hope all ye who enter here!

These mines produce the rare earth mineral neodymium. It is the most important element of the permanent magnet sitting astride each of the massive wind towers you see on your way to work, on TV and in journals. These towers are hailed as post-industrial art in the field. They are the direct cause of the death of thousands of Chinese men, women, and children each and every year. Shame on you who ignore these truths!

A 100 MW wind farm project's demand for neodymium cored permanent magnets will destroy nearly 40 acres of vegetation, leaving 6 million cubic meters of toxic waste, 33 million gallons of poisoned water, 600 million pounds of toxic tailings, and 50 tons of radioactive waste.[102] Externalities are not to be considered, of course, according to the wind industry.

We do not know the CO_2 impact from the work in Baotou province. Speculation would only lead to further disgust, so let's just ignore the increase in CO_2 emissions from this, the deadliest of Chinese provinces.

We also do not know the environmental impact of the constant O&M required—spare parts, lubricants, upgrades, new elements—to keep the turbines functioning well. We do not know the vehicle miles driven by the turbine attendants, engineers, managers, etc. We are not told the fuel consumed by the construction crews. Hundreds of cement trucks ply the access roads for each building site supplying the foundation components. We are ignoring the tons of rebar and other steel acting as the skeletal support for the towers. These negative externalities are ignored by all who participate in the discussion.

More environmental impacts must be considered, so hold on for a bumpy ride. When a concrete base is built to hold a thirty story tower upright in the strongest winds, it can consume two million pounds of concrete or 1,000 tons.

Let's do the math. Concrete is made of water, sand, aggregate, and Portland cement. According to the NRMCA, a concrete industry group, their 2008 report on concrete and CO_2 emissions,[103] the following statistics apply. They are quite similar to the figures from the EPA.

One ton of CO_2 is produced for each ton (2,000 lbs.) of cement produced.
45% of this is reabsorbed during the life of the concrete.
Approximately 10% of poured concrete is cement.
10% of this concrete is embedded CO_2 by weight.

A ton of poured concrete is 10% cement by weight. There are 1,000 tons of cement in two million pounds of poured base. This produced 550 tons of net CO_2 (55%). Ten tons of CO_2 are embedded in the concrete (10% of 10% of two million). Therefore 560 tons of CO_2 are produced from a two million pound concrete tower installation.

Place 100 of these concrete supports beneath 100 wind turbines and you begin to see the CO_2 results. Do this at 100 installations and you get some serious CO_2 fatigue for the planet. To be fair to the concrete industry, they have seen some improvement in CO_2 reduction from cement production since 2008.

For the sake of balance, let's listen to the AWEA which estimates 1.7 million cubic yards of concrete (170,000 cubic yards of cement) were used for wind power installations during 2009, providing the base and tower structure for 5,700 turbines.[104] That is 45,900 tons of cement.[105]

Let's do the math. How many tons of CO_2 were released from 45,900 tons of produced and poured concrete in 2009?

45,900 x 45 ÷ 2,000 = 1,032.75 tons of CO_2 were released in 2009.

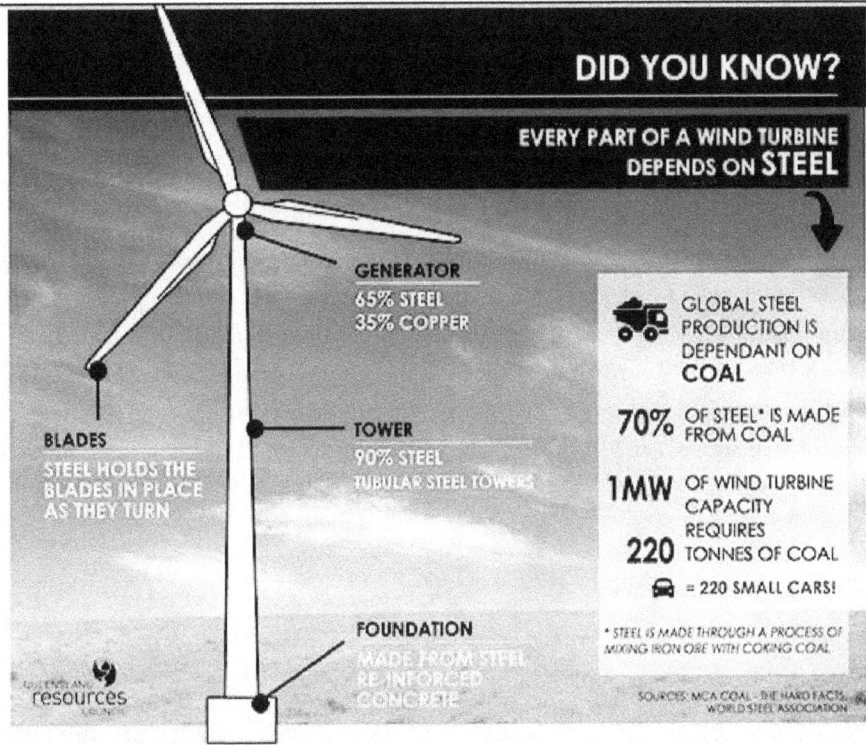

DID YOU KNOW?

EVERY PART OF A WIND TURBINE DEPENDS ON **STEEL**

GENERATOR
65% STEEL
35% COPPER

BLADES
STEEL HOLDS THE
BLADES IN PLACE
AS THEY TURN

TOWER
90% STEEL
TUBULAR STEEL TOWERS

FOUNDATION
MADE FROM STEEL
RE-INFORCED
CONCRETE

GLOBAL STEEL
PRODUCTION IS
DEPENDANT ON
COAL

70% OF STEEL* IS MADE FROM COAL

1MW OF WIND TURBINE CAPACITY REQUIRES

220 TONNES OF COAL

= 220 SMALL CARS!

* STEEL IS MADE THROUGH A PROCESS OF MIXING IRON ORE WITH COKING COAL

SOURCES: MCA COAL - THE HARD FACTS, WORLD STEEL ASSOCIATION

resources

How much CO2 is emitted in the construction of a wind turbine? Far more than is admitted by the industry. A turbine with 45 tons of rebar and 480 cubic meters of concrete has already emitted nearly 250 tons of CO2. Check out this video for the details: http://www.stopthesethings.com/2014/08/16/how-much-co2-gets-emitted-to-build-a-wind-turbine.

Figures from the mining, extraction, refining, and transportation are not included. The fiberglass blades weigh more than 1½ tons each, composed of resins and fiberglass (all petroleum-based products). The nacelle weighs more than 12½ tons, essentially more plastic (hydrocarbon-based forms). The core of the permanent magnet can weigh more than a ton.

The U.S. and Japan controlled most of the rare earth mining in 2002. A decade later, China controlled virtually all of it: 97%. The mining and refining of the rare earth elements neodymium and dysprosium is essential to the high efficiencies of turbine magnets. These neodymium/iron/boron alloy magnets are the strongest and most expensive magnets ever designed. More than 90% of this mining happens in China's Inner Mongolia. As Greenpeace's local representative, Jamie Choi says so poignantly,

> "There's not one step of the rare earth mining process that is not disastrous for the environment. Ores are being extracted by pumping acid into the ground, and then they are processed using more acid and chemicals."

Finally they are dumped into a vast tailing lake easily seen on Google Maps that is very poorly constructed and maintained. And throughout this process, large amounts of highly toxic acids, heavy metals and other chemicals are emitted into the air that people breathe, and leak into surface and ground water. Here is the story from the BBC in April of 2015: http://www.bbc.com/future/story/20150402-the-worst-place-on-earth. Villagers rely on this for irrigation of their crops and for drinking water.

Whenever we purchase products that contain rare earth metals, we are taking part in massive environmental degradation and the destruction of communities. The fact that the wind-turbine industry relies on neodymium, which even in legal factories has a catastrophic environmental impact, is an irony. It is a real dilemma for environmentalists who want to see the growth of the industry, but we have the responsibility to recognize environmental destruction that is being caused while making these wind turbines (http://dailymail.co.uk/home/moslive/article-1350811/In-China-true-cost-Britains-clean-green-wind-power-experiment-Pollution-disastrous-scale.html).

The planetary destruction begins with the mining and refining. The human cost in Baotou is staggering. Radiation levels around the tailings lake are ten times normal. A sample taken back to the U.K. with the April, 2015 BBC team was found to be three times more radioactive than permitted—in China. Nothing lives here: plants and animals are all dead. Humans survive for a few years but nearly all suffer from cancer, osteoporosis, skin diseases, and respiratory diseases. This is a boom town gone mad. Even *Huffington Post* is angered over this environmental catastrophe.[106]

A modest 3 MW turbine contains 1,200 pounds of neodymium and 130 pounds of dysprosium. Demand for neodymium is expected to grow by 700% by 2030 while that for dysprosium is expected to grow by as much as 2,600% over the same time period.

Enter the new capitalists in Beijing. They first attempted to corner the market in 2010 with control over 95% of global RE production. China tightened export controls by 40% and cut off the Japanese from their supply chain.

Funny things happen when you try to rig a market. Nearly half of Chinese exports to cut-off Japan went unregistered. New sources had opened around the world with the increased demand and Chinese supply restrictions. Turbine manufacturers reduced the amount of neodymium while increasing that of dysprosium in their core magnets. The global price for neodymium and dysprosium plummeted as much as 80% by 2012. A $17 billion a year market is now less than $1 billion. Julian Simon bests that Stanford fellow yet again.

While Chinese rare earth production has ramped up from 130,000 tons to 168,000 tons in 2015, it has failed to meet its own quota. Today the global shortfall is 40,000 tons as demand is accelerating from the wind industry. Meanwhile miners are injured. They sicken and die in the vast open pits and mines of Inner Mongolia.

Mining a ton of these materials produces nearly the same amount of radioactive waste according to The Institute for the Analysis of Global Security.[107]

In 2012, the U.S. installed a record 13,301 MW of capacity. If we average the figures for radioactive pollution from MIT and the Bulletin of Atomic Scientists (http://stopthesethings.com/2013/11/08/wind-power-the-poisoned-chalice), this figure is in excess of 4.5 million pounds of radioactive tailings produced to refine the rare earths required for these turbine magnets.

In comparison, the U.S. nuclear industry produces 4.4 million pounds of spent fuel each year. The nuclear industry provides us with nearly 20% of our power generation. Wind provides us with a scant 3%. Nuclear fuel management is a mature and security aware industry. In China, the tailings are dumped in an endlessly growing lake of horror.

Not only do rare earths create radioactive waste residue, but according to the Chinese Society for Rare Earths, "One ton of calcined rare earth ore generates 9,600 to 12,000 cubic meters (339,021 to 423,776 cubic feet) of waste gas containing dust concentrate, hydrofluoric acid, sulfur dioxide, sulfuric acid, and approximately 75 cubic meters (2,649 cubic feet) of acidic wastewater."

Rare earth minerals production has increased measurably since these 2009 figures were released. More turbines have been erected, consuming more concrete and neodymium, emitting more carbon dioxide, releasing yet more radioactive waste, killing more Chinese mine workers, and killing hundreds of thousands of birds and bats across the country. People near them are sickened by their endless thrumming noise.

We are told to be proud of these achievements as we march ever forward to the future of a carbon free world. We are also warned that global warming deniers must be punished.[108] 20 climate scientists have recently asked that such deniers, including fossil fuel companies and their supporters, be punished under the RICO Act for crimes of "knowingly deceiving the American people about the risks of climate change, to forestall America's response to climate change."

They are led by Jagdish Shukla of the IGES (Institute of Global Environment and Society). Does it matter that 98% of the institute's $65 million budget over the past 14 years comes directly from those government agencies that it supports with research? The Chairman of the House Committee to which the "RICO20" submitted their proposal responded,[109] "IGES appears to be almost fully funded by taxpayer money while simultaneously participating in partisan political activity by requesting a RICO investigation of companies and organizations that disagree with the Obama administration on climate change."

Does it matter that the EPA has a $700 million budget for weapons and armament and a PR budget of $15 million? Does it matter that it has become a government unto itself, writing laws and even issuing script in the form of permits to use or exploit?

The beat goes on.

CHAPTER EIGHT
PASSERINES

Enjoy the dance of the passerine flocks in this long video from Scotland. It highlights many of the small birds in the Uplands of the Esk Valley. Flocking is a defense mechanism against airborne hunters such as falcons and goshawks.

https://vimeo.com/27411516

We know passerines because they are all about us. The group makes up nearly 50% of all birds on the planet. 4,000 species of songbirds have vocal organs for communication, for territorial definition, and for mating. Calls range from the beauty of thrushes to the trying caws of ravens and crows. Size matters for small and large. Physical characteristics determine hunting area, food source, nest type, and location as well as a host of variables important only to the birds.

Their slaughter in turbine blades increases with their proliferation. Thousands of birds die each day during migration through the Death Blades. Two-thirds of all avian deaths from turbine encounters are songbirds. One quarter of a million die each year in North America from such collisions.[110]

This kingfisher emerges from the shallows with a meal for its brood. Hunting occupies the majority of every bird's day. Each consumes nearly it weight in food daily and brings smaller amounts to nests for chicks. Feeding of mates is often done while brooding eggs.

Wildlife

Now we descend into the heart of darkness. Some of this will be gruesome. There is no other choice. We must discuss the facts as they are presented. If you are offended by images of death, do not watch the videos or stills. They are real and graphic. The camera does not lie, unless made to do so by the imager.

Several issues arise regarding wildlife mortality by wind turbines. Among them:

Population impact
Phylogenic impacts
Site locales on migratory flyways
Migratory impacts
Nesting failures

A framework for analyzing environmental impacts of wind projects is obvious by its absence. It is missing in action. In 2007, The National Research Council suggested a framework, but it has yet to be seriously considered.[111] The USFWS has a wind turbine guidelines advisory committee. It has released a draft of recommendations, although they are yet to be reviewed or finalized. It is a lovely flow chart analysis of the proposed governmental decision tree. It defines words, methods of data collection, questions to be raised, and metrics of decision making. Meanwhile, birds die by the hundreds of thousands. The USFWS states clearly that "We lack a comprehensive estimate of avian or wildlife fatalities at wind projects. Having a better understanding of eagle mortality at wind projects will vastly improve our capability to develop advanced conservation and mitigation practices."[112] The killing has been going on for nearly a decade!

One source for frameworking this issue comes from a perfectly competent source, the Audubon Society. A less managed group of people would be hard to imagine and impossible to design! This august body of millions of friendly birders has been researching our avian friends for more than a century. Its annual Christmas

Bird Count Survey has documented the activities of hundreds of millions of birds for decades. It is the deepest and richest avian data source for any continent. Even our British cousins, no toshes when it comes to birding, are in awe of the simple work by thousands of volunteers over a two week period each winter.

From this simple source comes a straightforward suggestion for the framework of analysis for studies of any energy source's avian mortality impact. It is this pyramid: avoidance, minimization, mitigation.

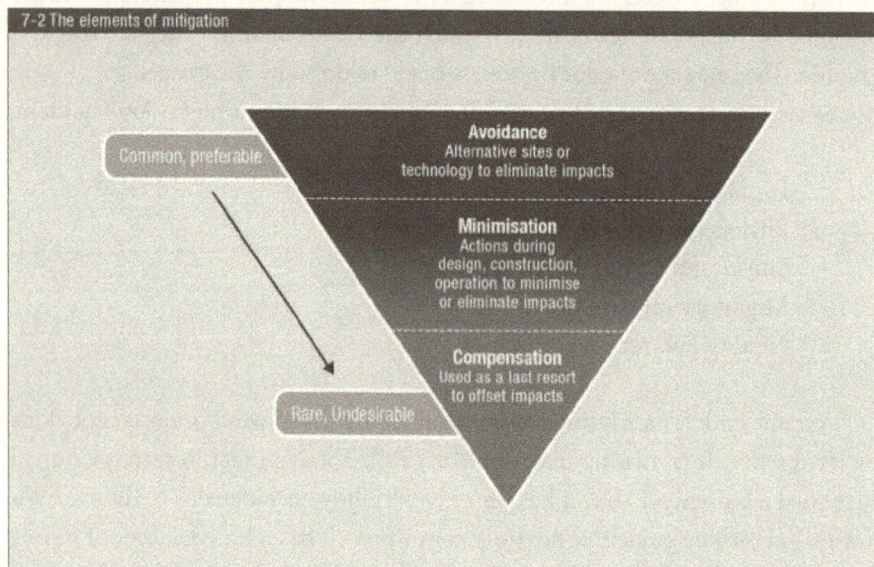

7-2 The elements of mitigation

Common, preferable

Avoidance
Alternative sites or technology to eliminate impacts

Minimisation
Actions during design, construction, operation to minimise or eliminate impacts

Compensation
Used as a last resort to offset impacts

Rare, Undesirable

An even simpler process is the following, suggested by Jesse Grantham, retired USFWS:

1. Identify all major and secondary flyways in the U.S. by region, state, and local level. These flyways are off limits for wind development.

2. Identify areas of high prevailing winds that are not in flyways (this information is readily available) and encourage development here.

3. Give highest priority to develop wind generators that minimize avian impacts.

4. Monitor for a minimum of 6 to 8 years each of the potential sites in all weather conditions on a consistent basis (daily) during all daylight hours with radar and qualified observers.

Here are two sets of easy techniques to answer the question, "Do these sites make sense?" Find the best site, monitor it prior to construction, use technology to reduce avian impact, and compensate where necessary. The order is important. Compensation is the least best, aka the worst, choice. Few in the industry would agree. Fewer in government would concur. The Aeolian capitalists seek returns upon invested capital—understood. When all falls to this boosterish Baal, it is incomprehensible.

These disturbing images haunt you night and day because you love these eagles. Some you find are headless and it always bothers you if you can never find the head. Others are cut in half and missing parts as well. Some you see are severely wounded, running around on the ground without a wing so that have to be caught so they can be put down or euthanized. Others have injuries you can't even see.

Here follows a series of quotes from an anonymous USFWS source, a technician, whose story should shock you.

In the beginning you were told that some birds would die but you were reminded that it wasn't that many, especially when compared to so many of the other things that kill birds. Besides the industry's new turbines rotate much slower and are much safer for birds. You were also told that this is an industry that serves the greater good of your community and the world.

As years roll by, you are promoted and given new responsibilities. One of them is to deal with the carcasses from around the turbines. While you were clearly told from the top that you are supposed to report them, you were also reminded from others in the company that once out in the field, you are alone. No one is really aware of what you see or do. Most of all, reporting everything will likely lead to wind farm shut downs.

With your new responsibilities of handling the carcasses, you realize that the carcass numbers are starting to add up. You are also seeing carcasses of beautiful strange birds that you had never seen before or even knew existed. So you start doing research to find out more about them. You discover the many of the birds you are finding are rare and are highly protected by strict laws.

You do additional research on wind turbines and start learning about their long history of killing birds. Your perception of wind energy is changing because of what you have learned and have experienced in the field. Turbines are sprouting up everywhere and you realize that the eagles and other species you thought were so protected are not.

There is great sadness from finding these carcasses, and especially the dead eagles. It pierces your soul and the images of the dead eagles you have found won't leave your brain.

These disturbing images haunt you night and day because you love these eagles. Some you find are headless and it always bothers you if you can never find the head. Others are cut in half and missing parts as well. Some you see are severely wounded, running around on the ground without a wing so that they have to be caught so they can be put down or euthanized. Others have injuries you can't even see.

One eagle in particular, with its last bit of energy, sank its talons deep into a piece of wood and you had trouble pulling the massive talons away. You think to yourself this poor guy fought to the very end. You look ahead with dread because there will always be a next eagle that you will have to deal with.

*There is also guilt because you are aware that these turbines are killers and when you tell people what you do, you think to yourself they might be aware that you work for the **eagle killers**. There is also guilt because you are supposed to report all the carcasses but if you report too many you will lose your job.*

You no longer see the greater good in wind energy and hate the job you were once so proud of because your job, your career, and your company are slaughtering off these eagles. There is no pride left in what you do, just shame.

But gathering data is neither easy nor quick. One has to spend long hours monitoring a location over many years, day after day, since we don't really know what triggers these mass migrations of birds on a given night, through all types of weather conditions. To do it right, we have no choice.

To their credit, the wind industry/regulators do have suggestions for consideration:

> Employ radar or human observers in reporting raptors.
> Shut down or slow blade rpms when observed.
> Place more open spacing between towers.
> Use a taller monopole tower design.
> Increase cut-in speed from 7 mph to 11 mph.[113]
> Be sure long-term siting reviews include migratory pathways.
> Increase research and conservation funding.

These ideas are fairly dated and rarely applied. Radar has been tried for nearly six years. Spacing is constantly being reduced. The monopole design is now ubiquitous. While its height takes it above much ground turbulence, it is murderous during passerine migration and breeding seasons. Heights are now being increased yet again "for better laminar airflow across blade surfaces." Increasing cut-in speed was suggested five years ago, yet the author can find no current reference to its use. Siting reviews that incorporate migratory flightways have yet to make financial sense, or much less been written. They entail multiple-year studies of site possibilities. These studies appear once the turbines are in operation, if at all. They are haphazardly done weeks after migratory deaths by low paid shills for the wind farms.

The overarching commentary from the industry, its regulators, and its *agit prop* lobbyists remains in debit:

Turbines kill far fewer birds than other energy sources, by several orders of magnitude.

More birds will die in the onslaught of climate change.

We must kill a few to help the remaining struggle.

These peevish views are less reflective of the real world than of their imagined Future World. In fact, were we to rescind our belief in global, human-induced climate change, the entire discussion becomes moot. Absent climate change, why kill the birds?

At a subsequent meeting this sad story unfolded about a man trapped by his conscience, his family obligations, and an unknown future.

The meeting lasted for about 1½ hours. It was more like an Act of Contrition with me being the priest. He said he was a Christian but had a belief system towards eagles similar to Native American people, a belief that all eagles have a special spiritual connection with the Great Creator and were specifically chosen to be the masters of the sky. He also believed that eagles deserve the highest respect because in this world they represent truth, strength, courage, wisdom, power, and freedom.

But as he has found out, wind energy and these turbines show eagles no respect.

The slower rotating blades fool these eagles into thinking they can fly through these openings in the blades. He said, "When they see prey across a hillside they just go for it and don't even think about the blades because they want that food. They tell you these big turbines are safer but I know the tip speeds are faster and it is not true."

While he was talking, I was shown dozens of eagle carcass images. I commented about how many of these images he had and that the industry does their best to keep them away from the public. He then added that the wind company had thousands of similar images.

The industry and its government advocates continue to remind us that hundreds of millions more birds are killed annually by buildings, windows, cars, and cats than by wind turbines. The figures vary significantly according to the source:

Cats:	365,000 – 1 billion
Buildings:	345 million – 1 billion
Powerlines:	100,000 – 170 million
Pesticides:	70 – 90 million
Cars:	60 – 80 million
Towers:	5 – 6 million
Turbines:	140,000 – 550,000[114]

This source is rather different:[115]

Cats:	1 billion
Buildings and windows:	580 million

High Tension Lines:	130 million
Pesticides:	60 million
Vehicles:	80 million
Wind Turbines:	negligible figure

Another source lists the deaths as a percentage:[116]

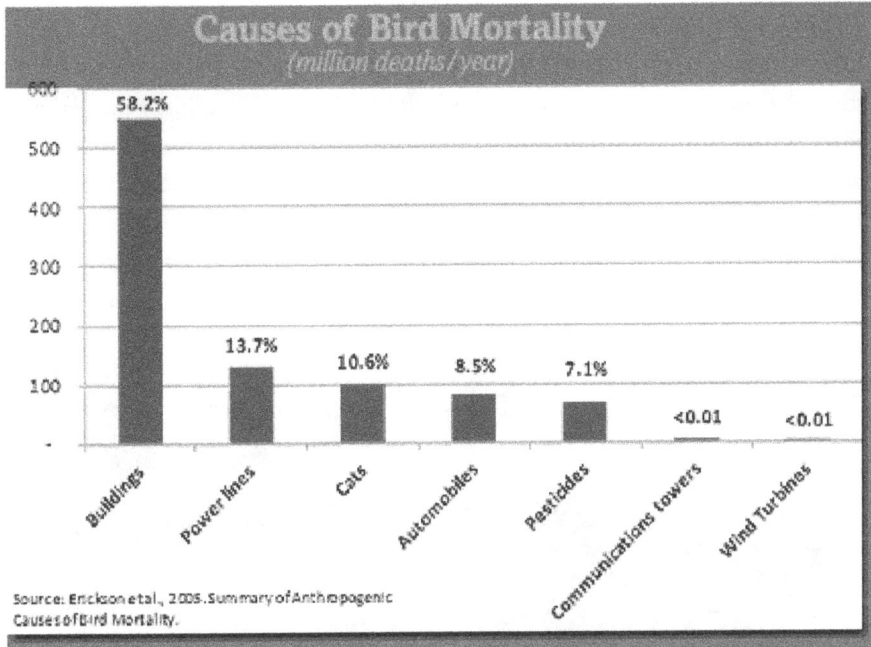

Causes of Bird Mortality
(million deaths/year)

Source: Erickson et al., 2005. Summary of Anthropogenic Causes of Bird Mortality.

These tables bear almost no resemblance to one another. The data sources are not quoted in the first two referencing texts. The British source does state that the figures are "highly uncertain." No foolin'.

Relative morality is a con job, as evidenced by these lists of relative mortality. It assumes your inability to interpret data. It assumes deaths of a bat, passerine, or condor are co-equal. It builds on the facetious argument of greater is worse.

One could as easily say, "More humans die in wars than on city streets; therefore murder in the streets is unimportant." The moral compass of comparable deaths is serpentine at best, obscenely illogical at worst.

The relativistic morality of the comparative deaths statement ignores...

Migratory flight paths confluence with wind farm siting,
Apex or hierarchic effects of raptor deaths on the food chain,
Species impact vs. total estimated annual bird deaths,
Nesting failures: 100% probable when one parent dies, and
Species wide *multi-generational effects* of bird kills.

Sites for wind farms are often on hillsides and mountain passes in remote areas, particularly in the Rockies and along the Western littoral. These sites also reflect the *migratory patterns* of avian transitory lifestyles. Confluence is perhaps the most dangerous intercept between avian and human species. Raptors and their prey can migrate for thousands of miles. The flightways coincide neatly with ideal turbine site evaluations.

Grab a hand full of sites and look at their potential impacts on migratory birds. You stagger backward with the overwhelming magnitude of the slaughter.

Literally tens of thousands of birds move north each spring, from mid-March to mid-May. They reverse the flight in the fall. Turbines stand in the way. Thousands of turbines. For hundreds of miles the flocks must run a gauntlet impossible to imagine. Birds, thousands of them. Eight species of terns, laughing gulls, three species of ibis, teal, flocks of eastern kingbirds and scissor-tailed flycatchers, foraging pelagic birds like gannets and jaegers, thousands of shorebirds, some heading as far north as Alaska. It is one of the natural wonders of the world.

Don't stop here. Move inland to the prairies of the Midwest, the Appalachian mountains of the eastern U.S., the Atlantic Ocean (where the recently federally restored Atlantic puffins now winter), to the western U.S. where passerine migrants are met by a massive sets of spinning turbines as they funnel up narrow canyons.

The slaughter does not go unnoticed.

Sekano.Net

The hunters and carrion eaters take up residence at or on turbines without knowing these are death traps.

As we saw with the opening images from Crete, the lazy gyres of a buzzard or griffon beckon others to the feast below. The griffon flight can rapidly become a *dans macabre* if it is hit by a rapidly rotating blade. The dying giant raptor becomes yet more food for the remains of the kettle still forming skyward. The birds' eyes are naturally keened to the ground, ignorant of the spinning blades. Even the fastest cannot get between the spinning blades. Blade tip speed can exceed 170 mph, with a blade interval of less than 2 seconds. The best fighter pilots couldn't make the entry and exit unharmed.

Raptors fly between ground level and 300 feet. Rotor sweep for today's 2 MW turbines crosses 75 feet to 275 feet. The flight zones converge. Raptors use the same flyways as other avian migratory species. Turbines site along these identical flyways. The convergence is purposeful for the operators: Go where the wind blows steady. Kills increase during the fall when flights are more frequent, just as young inexperienced birds leave the nest.

Apex killings. Raptors occupy the top of their food chain, much as humans do. The effect down chain of an immoderate number of deaths can have a multiplying result. Call it the *death ripple*. While the number of raptor deaths may be relatively few, they have an overarching impact upon other species upon which they feed or with which may share food resources. Wind turbines currently kill nearly 90,000 raptors each year in the U.S., or 18% of all such bird deaths (songbirds, passerines, are 60% of all turbine driven deaths).[117] These figures will rise as the number of turbines increases.

Since 1997 over 31,000 unaccounted for eagle carcasses have been shipped to the Interior Department's Denver Eagle Repository. How many more eagles killed by wind turbines were never reported or shipped?

The DOE expects a 12-fold increase in the number and size of wind farms by 2030. This increase will be more than numerical—it will have a multiplier effect. Let's do the math:

U.S. electricity production from wind:	3.5%
2030 goal for U.S. production:	20%
Raptor deaths today:	83,000
Raptor deaths in 2030:	474,286

The deaths of tens of thousands of raptors each year may average out to a meaningless figure. This homeopathic dilution to an inconsequential figure denies the underlying importance of the deaths of small numbers of highly important species. It is worth less than a measurement of the *species specific impact* of these deaths. Every species can enter a death spiral from which recovery is difficult. If a species takes years to mature, raises a single brood every other year, and lives upon carrion or hunting, it is more at risk to smaller death counts.

In the carrion-eaters case, their nest is maintained by the female for 3 to 4 months, protecting the chick. She rarely departs during incubation and the first two months of life—only if the male has returned and then for very brief periods. She and the chick are entirely dependent upon the male for survival. If he dies at the blade of a turbine, she dies, the chick dies.[118]

Adolescence lasts up to five years for raptors (50% of all eagles die within the first three years). They give birth to one chick each year or every other year. They reproduce for at most 35 years. Remove an *annually increasing number* each year and the results will cascade down the generations. If Rachel Carson could worry about DDT effects upon egg shell thickness, we must wisely worry about *multi-generational* death spirals in the raptor genus.

The wind industry and its supporters tell us that many more birds die from feral cats and tall buildings than from wind turbines. The numbers are significantly higher, certainly. The question not asked is what happens at the top of the food chain. As for passerines, what carcass remains after being shattered by the Avatar blades of a turbine?

Raptor populations are especially slow to reproduce. In fact, the *cat predation* and *building collision* problems are far more significant and dangerous to avian *passerine* populations. These are the songbirds of our world, the birds who bring us the beauty of the forest, the glen, and the river marsh. 1.3 billion to 4 billion birds, primarily passerines, are killed each year by cats in the U.S. The majority of these deaths are from un-owned feral cats, but home based friendlies are nearly as responsible for passerine mortality.

Perhaps a commission could be appointed to study the metrics of decision making at the feline/avian interface. Perhaps a *take limit* should be imposed upon our fiendish feline friendly killers.

Or perhaps this comparison is nonsense. Relativistic morality is easy to conjure up, more difficult to put back in the bottle. Be careful what you ask for, you might just get it.

Building collisions take another 600 million birds each year in the U.S. Again, the significant majority are small songbirds. Warblers, buntings, thrushes, including those that are endangered migratory birds, are most at risk. While many data sets are from the northeast, many authors assume similar mortality characteristics across the nation. The authors of two recent reports, Loss, Will and Marra, are the same authors we shall refer to regarding avian mortality. Many thanks to these men for their scientific research.

This *large scale, small bird mortality* (warblers, buntings, orioles, tanagers, grosbeaks, thrushes, and even hummingbirds, to name a few) is to be compared to *small scale, large bird deaths* at the top of the food chain. The argument is thus: the number of deaths is huge; a few more don't impact the overall population to any great extent. This is specious, at best. It is devious, at worst.

You'll notice that seldom is the word *cumulative losses* discussed. Cumulative losses are the overall losses for all mortality factors, both man-caused and natural. Nature has evolved to cope with *natural mortality factors:* changes in local weather conditions, disease outbreaks, local changes in habitat quality and availability, increased predation, differing food sources. Throw in man-caused mortality factors and wild animal populations find it difficult to adapt. Birds either have to start laying more eggs to compensate for the increased losses or the population is going to decline drastically.

Shall we review just a few of the dozens of information sources on avian fatalities? We shall try to ignore the more volatile websites, but this is a very emotional issue for many. Those who live near wind farms are the most concerned and agitated by the destruction of our avian populations. The following is a very brief summary of a few of these concerns.

The *Encyclopedia of the Earth* writes: "The sheer volume of bird kill does not begin to depict the magnitude of ecological damage, since the most susceptible species tend to be those which are keystone species or species already threatened by other human pressures. Additionally, bird mortality due to large wind farms is exacerbated by inherent linkages between bird behavior and wind farm siting decisions. Proponents of large scale wind farms (including some federal agencies), for example, tend to favor sparsely vegetated saddles or other funnel-like landforms, which are highly correlated with high density bird migration routes or raptor soaring locations."[119]

This is from LiveScience.com, quite the environmentally friendly website. No hydrocarbon aficionados here, just hard green facts. The story references turbine blade speeds and their killing fields.

Although they're turning at a slow, almost relaxed pace, wind-turbine blades actually move very rapidly: The outer tips of some turbines' blades can reach speeds of 170 mph and can easily slice off an eagle's wing. And when hawks, falcons and eagles are flying, they're usually looking down at the ground for prey, not glancing up to watch for a knifelike blade whipping down on them from above.

"There is nothing in the evolution of eagles that would come near to describing a wind turbine," says Grainger Hunt, a raptor specialist with the Peregrine Fund. "There has never been an opportunity to adapt to that sort of threat."[120]

Save the Eagles International offered this joint new release in February, 2012 along with their co-sponsor *World Council for Nature*:[121]

In 2012, new construction of wind farm turbines at the Tehachapi Pass forced U.S. Fish & Wildlife Service biologists to alert Kern County, California officials to the fact that most of the proposed wind projects, as well as at least one existing wind farm, are a threat to the condor. Nevertheless, county officials approved the sites for the lethal wind turbines where condors fly.

"Condors can travel 200 miles in a day," said Jesse Grantham, the retired California condor coordinator for the Fish & Wildlife Service, "as the bird forages for food or takes a road trip on a whim to satisfy its curiosity. The condor has evolved to be observant of and attracted to novel objects and activity as it must constantly scour vast landscapes for its dinner."

Adds Mark Duchamp, president of STEI: "In view of this, it doesn't take a rocket scientist to predict they will be attracted to wind turbines, and die in their arms as do golden eagles. Dead birds under the turbines will be another fatal attraction."

Vultures and other raptors perch on wind turbines. Condors will be likely to perch as well, and accidents will inevitably happen. Even if they did not attempt to perch, they are likely to get struck while looking down for food. The video from Crete can attest to this!

Notice the conflation of a long time career FWS employee with a radical conservationist. Both express similar views without reference or respect for one another. The story is the same from most sources—except the industry, its pet NGOs, and its lobbyists.

Birds threatened with extinction are at risk from wind turbines: condors, golden eagles, and bald eagles. These are all protected by Federal law. Except for when they are killed by Franken Towers, in which case their deaths are for a higher cause, the survival of the planet. Permits are issued for 30 years to assuage the capitalist recidivists that their investment is assured (Clearly they have never heard of market risk). Do what you can to help the birds, but don't allow their deaths to interfere with your salvation efforts.

This comes from our European friends concerned about avian mortality from wind energy sourcing. The stories from Europe are far too numerous to list here, so allow these few to be representative. Know that there are more than 2,000 private and public organizations fighting against the European Commission and the local governments to stop wind power aggrandizement.[122]

In Scotland, permits for the Edinbane Wind Farm in 2007 were temporally halted. The estimated kill rate for the golden eagle for 25 years of operation was 137, well in excess of survival rates for the rapidly declining species. The result? The developer changed their computer model to show 25 deaths and received their build permit. The collusion between the developer and the authorities was questioned, but to no avail.

Many, many more quotes could be added from the European experience. Much of Spain, Italy, and Greece have been negatively impacted by wind turbine avian mortality. The Spanish acceptance of wind farming exceeds even our own, to these consequences:

On 12 January 2012, at the First Scientific Congress on Wind Energy and Wildlife Conservation in Jerez de la Frontera, Spain, the Spanish Society of Ornithology (SEO/Birdlife) *made public its estimate that, yearly, Spain's 18,000 wind turbines may be killing 6 million birds and bats. The average per turbine kill ratio is 333 deaths per year, a far cry from the 2 to 4 birds claimed by the AWEA.*[123]

While we offer two quotes from STEI, we also note that the organization may represent the extremes of the environmental movement. Some of the more formal bird societies have banned the president from meetings. As we are quoting from AWEA on the one hand, the least we can do is include its antithesis.

Jim Wiegand, an independent wildlife expert and vice president of STEI, writes in *East County Magazine* (San Diego, CA) in January, 2014:

> *The golden eagle is a species in rapid decline. The primary reason for their population crash has been the development of wind energy in its habitat.*
>
> *There is a history of eagle nesting failures and of habitat abandonment near wind projects. This is a pattern that has been hidden from the public as wind projects have expanded across California and the west.*
>
> *These undisclosed impacts occur when adult eagles are killed by a turbine during the egg and downy stages of a nesting cycle. During this critical 8-9 week period there is a 100 percent probability of a complete nest failure if one adult eagle is lost. It simply is not possible for a single parent to hunt, incubate or protect their young from the elements.*
>
> *In 2012 the BLM in California contracted BioResource Consultants, Inc. to collect new field data on the current breeding status of golden eagles in California, to document golden eagle habitat use and population demographics. The primary study areas were approximately 2/3 of California.*
>
> *71 golden eagle nest sites were claimed to be active. Only 45 of these occupied nest sites were successful in raising their young. This is the important number for this declining population.*
>
> *A recent FWS estimate is that the CA wind industry's turbines are killing about 100 golden eagles each year. In this region there are the four wind resource areas, Altamont Pass, Montezuma hills, Tehachapi and Pacheco Pass.*
>
> *When studies are conducted, search areas are greatly restricted. Search intervals have allowed mortally wounded eagles up to 90 days wander off, be picked up by lease holders or wind personnel, or carted off by scavengers.*
>
> *The FWS currently estimates that the golden eagle population in California is about 2,000 birds. Yet the nest site data from the BLM surveys indicates fewer than 500 golden eagles.* [124]

As the data we've chosen appears accurate, but the tone shrill, I have significantly edited the comments here. You can read the entirety of the transcripts at the web link. The shrill attitude reflects a deep concern for the abeyance of law with regards to endangered species. Birders can get that way.

We have read evidence of site locale mistakes, nesting failures, demographic errors, mass European avian killings, and poor data collection. These are only the beginning of the problem.

USFWS retired personnel report that golden and bald eagle takes are increasing in most areas, while declining significantly in others. Why? The golden eagle population is approaching extinction.[125] The USFWS offers free shipping for all golden and bald eagles takes. The number of takes from each reporting period are clearly stated.[126] These figures, 3,868 for the first three quarters of 2014, increase with time. The new requests received is 4,350!

The regions with increasing requests coincide with turbine siting increases. Jim Wiegand estimates 31,000 have been recorded by USFWS since 1997. The Agency suggests users ask for the large size box that holds four to five carcasses, unless a more cavernous unit is required. "The Repository is happy to provide a prepaid shipping label via email."[127] Their purpose in holding the tens of thousands of eagle carcasses? Native American religious requests, of course. The number of requests honored? Exactly one, after the Supreme Court ordered it. This only took ten years.

Let's look at the report in California from the utilities Wildlife Reporting Response System (WRRS). This is their fatal reporting system into which all deaths are recorded.

Results

Unadjusted Fatalities

For bird years 2005–2010, fatalities documented during regular searches comprised 71 avian species, 14 of which were raptors (Table 3-1). Four species were nonnative species, including the two most commonly detected fatalities, rock pigeon (n=1,125) and European starling (n=619). Over 39% of fatalities detected were of nonnative species. The most commonly detected fatalities of native species included western meadowlark (n=524), red-tailed hawk (n=394), burrowing owl (n=278) and American kestrel (n=199).

Table 3-1. Annual Fatality Detections in the APWRA by Species, Bird Years 2005–2010

Species/Category	Bird Year						
	2005	2006	2007	2008	2009	2010	Total
American kestrel	20	47	51	35	29	17	199
Burrowing owl	28	126	49	22	36	17	278
Golden eagle	15	36	19	12	13	10	105

Now let's look at the reported fatalities for the same time frame from the USFWS:

Table 3-9. Annual Estimated Total APWRA-Wide Focal Species Fatalities (95% CI) Based on QAQC Detection Probabilities, Bird Years 2005–2010

Species/Category	Bird Year and Fatalities (95% CI)					
	2005	2006	2007	2008	2009	2010
American kestrel	207 (182–231)	344 (308–380)	347 (316–379)	213 (196–230)	203 (186–220)	205 (173–273)
Burrowing owl	237 (214–260)	837 (762–912)	285 (260–310)	138 (122–154)	216 (196–237)	190 (162–218)
Golden eagle	64 (60–67)	99 (94–105)	43 (41–44)	29 (28–29)	43 (41–45)	38 (33–43)
Red-tailed hawk	339 (322–357)	429 (405–452)	202 (195–210)	103 (99–108)	94 (89–100)	206 (191–220)
Total focal species	846 (778–914)	1,709 (1,568–1,849)	877 (811–943)	483 (445–521)	556 (511–601)	638 (559–718)

Somebody is cooking the books. Ten golden eagle deaths in 2010 from the first chart while the second shows 38 for the same period as recorded by UWFWS. Either APWRA and WRRS are false or USFWS is baiting us. These figures are illogical. Are counts being missed? Are they mixed with desk counts by formula? Are eagles being removed each day by staff? Are the counts 'adjusted'? How were the carcasses surveyed? Were they surveyed?

Yet, as of mid-2015, the Service is reporting an amazing drop in eagle mortality figures for Altamont. Here are the eagle mortality figures for Altamont as reported by the USFWS:

February 2013: 4   August 2013: 5
March 2013: 3   September 2013: 5
April 2013: 3   October 2013: 6
May 2013: 1   November 2013: 1
June 2013: 3   December 2013: 0
July 2013: 2

Total USFWS reported eagle carcasses at Altamont 2013: 33

January 2014: 1	July 2014: 3
February 2014: 0	August 2014: 2
March 2014: 4	September 2014: 5
April 2014: 3	October 2014: 4
May 2014: 5	November 2014: 2
June 2014: 8	December 2014: 0
July 2014: 3	

Total USFWS reported eagle carcasses at Altamont 2014: 37

January 2015: 0	May 2015: 0
February 2015: 2	June 2015: 1
March 2015: 5	July 2015: 1
April 2015: 3	August 2015: 3

Total USFWS reported eagle carcasses at Altamont through 8/15: 15

Are there so few birds left to kill that their mortality figures are declining? Are these recent figures so doctored as to be unfit for data consumption? Or, have they finally figured out how to shoo the birds away from their own avian hunting grounds?

Reader, you will have to decide the answer on your own. Perhaps you can ask USFWS their effective strategy for eagle shoo? Or, you can dive into the APWRA from 2010 by Leslie, Swartz and Karas.[128] This is the definitive report on the Altamont site, the largest grouping of turbines in the U.S. What stands out from this report?

> A review of search techniques, both physical and Bayesian,
> carcass counts and adjustments,
> maintenance staff interference and removal, and
> mortality rates impacted by season and shutdowns.

These are older data but they demonstrate the changes in behavior by turbine operators and search personnel. For example, 50 meter search patterns crisscross the turbine fields. This appears reasonable until you understand that large size carcasses fly further than 50 meters and that injured birds crawl away from the crash zone. Less rigorous searches than Altamont may wait weeks between events, leaving carcasses to scavengers and staff. Incidences are reported of eagle bodies found stuffed under rocks and in ground squirrel holes. In total 1,261 raptor carcasses were found, with 4,658 as the total body count for the report period.

These statements, along with the serious scientific studies from Loss, Will and Marra that we will examine shortly are *evidence specimens* of the mounting concern throughout much of the bird world.

Many large and formal organized bird groups are in support of wind energy. They take offense at some of these comments. Why, if the worries remain genuine, are these bird groups offended? The issues of turbine sitings in flight paths, species kills, apex impact, nesting failures, and intergenerational results of current kills are very serious. Many non-government organizations (NGOs) put their green eyeshades on when viewing avian mortality.

We have discussed significant evidence of collusion, obfuscation, and lies within the regulatory agencies, NGOs, and the wind industry. More is to follow. The public is shut out. The taxpayer is obliged to support this with hard-earned capital. The media is enamored by alternative energy projects, to the exclusion of fact-based evidence of bird deaths. Politicians—they are cheaper by the dozen.

500,000+ avian deaths each year means that within three years 1.5 million birds may die. In five years, nearly 3 million will die. In 10 years, 6 million or more will die. This is simply a straight line addition of each year's estimated deaths. No reference is made to the increases in turbines, in their size, and in their siting patterns along migratory flightways. No reference made to the 12x multiplier of turbines through 2030 and their effect on avian mortality. This is simply adding the current death list.

How long can the raptors hold out against such slaughter? When will extinction be the final solution? With no more birds to kill, will we finally be free of the problem?

In a more cynical world you could image a scenario. Developers and regulators are sitting around a table bemoaning the restrictions of the Migratory Bird Treaty Act of 1918. They are complaining how birds in general are a problem:

How can we rid this country of those pesky birds which get in the way of all of our plans? Certainly plate glass windows are having an impact but they are not taking enough. Feral and domestic cats are contributing, but not enough to significantly dent the populations. Pesticides and poisons are hit and miss; they are impacting a smaller segment of the population. And loss of habitat is having an impact but not quickly enough. We need something fast acting that will take out large numbers of birds in a short period of time.

We need to go after large numbers of birds at one time that are either concentrated at feeding sites or in migratory flyways. Kill hundreds, even thousands, at one fell swoop, and do it consistently over a period of a number of years, and do it under the guise of saving humanity or green sustainability.

Who would dispute this? Even the bird and conservation groups couldn't or wouldn't protest. O.K., sounds good. What would we use? Someone else suggests. Turbines with huge 450 foot whirling blades, back to back, a veritable net miles wide, that will kill large numbers of birds on nights when they are concentrated on spring or fall migrations, when they are unable to see the spinning blades, on overcast or foggy nights.

And the greatest fact of all: The dead birds will be virtually undetected, as the spinning blades will annihilate any evidence. Smash them. I'm sure we can find consultants and government agencies that will carry some kind of bogus credentials who will look the other way.

We can certainly get federal agencies to pay us to build these towers. The IRS will give us tax credits. The EPA will give us research grants. The DOJ will give us legal cover. The DOI & FWS will give us regulatory cover. Wall Street will give us the little capital we actually need.

The irony is that in almost every case birders and biologists familiar with these areas (via Audubon chapters that have been watching birds for years and submitting their observations to their chapter minutes and field notes) have raised a call for concern. But, since they have not gathered the data in a systematic way that fits scientific standards, their data is considered useless, even though reports have been gathered for half a century or more.

Sam also warned me: "Every time a writer, journalist, or reporter contacts ANY USFWS agent about a wildlife investigative issue, that agent has to immediately seek approval after the bureaucrats in D.C. get the details and the nature of the query."

In order to find the truth about wind farm mortality one must first get past the industry's confidentially or nondisclosure agreements. It seems everyone connected to the wind industry must sign them including researchers. As a result of these agreements, nothing stated by these people can ever be trusted to be true because information is being filtered.

It is a disgrace that the wind industry researchers and wind energy employees have to sign confidentiality agreements. This is not the CIA. It is a highly profitable, but rinky-dink source of energy killing off precious wildlife species. But most importantly with these confidentiality agreements, there is no science because truthfulness and objectivity is lost. How can the truth be disclosed when what you say has to meet with the approval of your employer? How can the truth be told when the true scientist cannot even design their own studies?

Contrast the imaginary conversation above with the responses from these same groups (regulators, the wind industry, and many environmental groups) regarding fracking. Hydraulic fracturing has been going on since 1948. More than 2.2 million wells have been fracked without a single incidence of water table infringement from a well drilling operation. Not one well out of more than two million. Even Lisa Jackson of the EPA has testified to this before Congress. Yet the *fear of possible outcomes* drives much political decision making.

Attending a recent frack meeting for the public in California, I listened to the tale of "threats to the Sacred Waters, the air, to Mother Earth Herself." Despite the facts (no water table infringement, vast improvement in air quality from CO_2 reductions, and an esteem for the earth evidenced by every oil and gas worker and executive I have met), fear of the unknown was the most important consideration. "But what if…?"

The wind industry and its supporters tell us that many more birds die from feral cats and tall buildings than from wind turbines. The numbers are significantly higher, certainly. A question not asked is "What happens at the top of the food chain?" Raptor populations are slow to reproduce. Nesting interruptions can be 100% fatal to the chicks. Population density is far lower for large birds. Life expectancy is longer, so adolescence is as well.

Do the birds have some say in this? They either have to start laying more eggs and producing more chicks to compensate for the increased losses or the population is going to decline drastically. Laying more eggs and producing more chicks is not going to happen. Extinction will happen.

Raptor deaths have a geometrically negative impact upon their local ecosystems by allowing food sources to multiply. This reduces wetlands and encourages habitat destruction by rabbits, deer, and others. The long term effects of such depredation are clear from previous soviet attempts at top down management—the Aral Sea, the Caspian, TVA, and others. Ignorance of nature leads to its destruction.

Today passerine deaths are far greater in number and in frequency. A visit to any of the major bird migratory *flyways* in the spring or fall puts into perspective how incredibly effective the massive killing effort has become.

In south coastal Texas huge flights of raptors, shorebirds, and passerine migrants ride the strong southeasterly winds pushing up out of the Caribbean in the spring, winds that can push them all the way to their breeding grounds in the Plain States, Canada, and Alaska. They move across scrub and city, across hills and freeways. Weather doesn't often slow their progress. The biological urge to nest, to mate, to carry the species forward is inescapable. These small songbirds cover thousands of miles. They may travel at night. They may nest in trees or scrub in the afternoon. They feast upon the insects and worms of their natural lifestyle.

In previous years a Director of Land Management for the National Audubon Society in Texas had seen thick ribbons of raptors, shorebirds, and passerine migrants passing overhead on early spring mornings. These ribbons of birds stretched for miles ahead and behind, traveling the wind currents felt only by them. There were tens of thousands of birds, perhaps millions, moving up the coast each spring. They passed overhead during the day and at night. They did so in inclement weather and under fair skies.

Today, these avian storm clouds have become a drizzle of birds. Excessive human mortality factors—pesticide poisoning, destruction and alteration of native habitats, shooting, and wind turbines—are the killing machines. Texas has built massive wedges of turbines directly across the flyways of these avian miracle workers. Why? The same wind currents traversed by the birds are the ideal site for turbines. One could not have designed a more efficient multi-species killing machine than the wind turbine.

The passerines survive all but the ubiquitous blade. Nocturnal passerine flights occur at greater elevations, typically 400 feet.[129] They fly directly into the vision obscured blades of the turbines. Passerines endure the largest killings of all bird types by wind turbine blades, 60 to 70% of all avian mortality for this shrinking genus.[130]

Passerine deaths exceed one million each year, but the estimate is unsubstantiated. Why? Turbine operators do not allow outside biologists to perform carcass counts. Their counts are formulaic and intermittent. They waste years discussing which formula is better at estimating death.

Wyoming is the newest, windy state, for turbine towers. 1,000 have been approved by the Interior Department. The federal friendlies expect 46-64 eagle takes each year. A few miles down the road, Duke Energy has 110 turbines killing five eagles each year. It has the claim to be the deadliest Duke. Of their 15 wind farms, more birds die here than at any other of their facilities.

The company does have an active policy to try to reduce avian deaths. They have tried to site the towers wisely. They remove carcasses that may be food for other raptors. They move rock piles which the nesting pairs could use for homes. They experiment with radar and loud noises. They have shut down turbines during migration. This has had some measure of success: no deaths during the shutdown. When an eagle dies, no one can remove the body except a federal employee of the USFWS; the body is covered with a tarp to avoid predation form above, awaiting the undertaker. Duke prides itself on good stewardship.

Good conservation begins with conserving rather than destroying. Perhaps the USFWS and AWEA could take a few lessons from an evil corporate giant. Yet Duke was fined $1 million for a dozen avian deaths in 2012.

Meanwhile, back in energy purgatory, BP suffered the most egregious federal wildlife fine in history: $100 million for deaths, real and supposed, from the Macondo Gulf of Mexico oil rig explosion in 2010. Confirmed avian deaths were 6,187 from the explosion and aftermath.

In 2009 PacifiCorp, a powerline company, was fined $10.5 million for similar deaths of 232 eagles on its transmission lines. The estimated 20 eagles that died from turbine death at the same company's wind farms have not required monetary restitution. Duke has paid the only fine ever collected against a wind energy firm.

Let's do the math:

BP (Oil and Gas): $100 million mortality fine / 6,187 confirmed bird deaths = $145,560/corpse.

Duke (Wind): $1 million mortality fine / 12 confirmed deaths = $83,333/corpse.

PacifiCorp (Power lines): $10.5 million mortality fine / 232 confirmed deaths = $45,258/corpse.

It seems the price of death varies with the perpetrator. Some are more guilty than others. Orwell would be ashamed that such a statistic could be compiled, much less compared. My apologies to the dead.

Petro death is a punishable offense; blenderizing is fine. Move right along. Nothing to see here. Mind your business. Have a nice day.

Do you see a pattern of broken promises? Three separate acts of Congress define the treatment for at risk avian deaths of endangered species. Punishment is doled out only to firms *not* associated with alternative energy sourcing.

Before we leave this cesspool, you may want to hear of the 2010 settlement in California between a group of turbine operators and court petitioners:[131] $10,500.00 per MW of installed capacity for "each phase of repowering... to be paid in three annual installments." The petitioners were local Audubon Societies. Funds are to be paid into their local park districts, research (PIER), and conservation plans. The 480 MW of installed capacity will generate over $5 million in mitigation fees. Plus the $300,000 annually for monitoring. Sounds like Blood and Oil.

We have seen how this plays out. Money for blood. Retribution and retaliation. Applied to wind turbine operators, the cost is much lower, certainly. We can't do the math because we have no idea how many birds were killed by the Franken Towers. We do know that these same operators were, in April of 2015, given permits to run their facilities despite clear evidence of the avian slaughter.[132] They protested that any other solution would drive them to bankruptcy.

Poor darlins'. Shame about the birds, but we gotta make a livin' here.

CHAPTER NINE
WHOOPING CRANES

https://www.youtube.com/watch?v=DGX52B9iXXU

Passerines and raptors are the most obvious deaths in the turbine killing fields, but other birds suffer as well. The whooping crane is an endangered species native to North America. Its population is well tracked in the eastern U.S., with annual counts of births, deaths, and causes of death noted.[133] Meanwhile, the Midwest and western areas have almost no tracking. The Aransas-Wood Buffalo population (AWBP) of whooping cranes, the last wild migratory flock of this species, is holding steady.[134]

As thousands of new turbines are erected up and down the Central Flyway, this small flock of cranes is at risk. The leading authority on the whooping crane, Tom

Stehn, retired USFWS, has said, "It is important to analyze the potential impact of literally tens of thousands of wind turbines placed in the whooping crane migration corridor in the coming years."[135]

Incidental Take Permit: A License to Kill

What is the industry and USFWS response? USFWS is considering issuing the first-ever Incidental Take Permit (ITP) to a wind farm for the killing of endangered whooping cranes. The U.S. Fish and Wildlife Service is acting at the request of the Wind Energy Whooping Crane Action Group (WEWAG), a collection of 19 of the largest wind energy developers.[136] The American Bird Conservancy and 75 other conservancy groups have called for a full Environmental Impact Study (EIS), on the proposed 100-turbine project that would be built in a key migratory pathway for many birds in the Prairie Pothole Region of North Dakota.

The AWBP of whooping cranes is likely to be adversely affected, as the entire population migrates through the area in the spring and fall. USFWS itself has already stated that even a 3% annual mortality factor in excess of the norm for the flock (8 birds) would destroy the species. The "mortality of any birds in such a small population represents a loss of genetic material and a setback for recovery efforts."[137]

George Archibald is the co-founder of ICF, the International Crane Foundation. You may remember him from the *Tonight Show with Johnny Carson*. He had Rex, a female whooping crane, with him then. These are his thoughts on the turbine-crane interface. They are thoughtful and intelligent: "I'm concerned about the placement of wind turbines. We know that whooping cranes on migration roost at night in small wetlands, little ponds, and that they feed in nearby cornfields or wheat fields. If you put wind turbines in these areas, it could be much more dangerous. We're all for clean energy, but on the other side you have to be careful. I think there are compromises."[138]

This is Jesse Grantham from the Condor Project: "If you look at cranes, they can breed at 4 to 5 years of age, breed every year, and will produce a chick every year. (Cranes lay two eggs and generally hatch two chicks, but the stronger chick usually kills, or drives off the younger weaker chick. Called Cane and Able syndrome). Cranes are very dependent on ideal breeding habitat for chick survival, which means proper water levels in marshes where they breed. Weather conditions play a major role here. Same goes for wintering habitat. Now throw in some human caused mortality factors, shooting, and collisions with power lines, and the bird takes a big hit, Again human caused mortality will be indiscriminate, taking adults or immatures."

Compare the comments from these birdmen to those of the regulatory world in Nebraska. Let's use the Cornhusker State as an examination area to dig deeper into the processes and suggestions brought forth by the regulatory and environmental community.

The challenges facing the whooping crane are significant, yet correct responses are largely unknown. With good foresight and care, deaths can be avoided. Is this evident in the actions of USFWS? How do the actions of WEWAG compare to Archibald's commentary? Time will tell.

Their mission statement is long on renewable energy commentary, short on extinction prevention.[139] They seek regional, multi-species take permits for a list of endangered species, foremost of which is the whooping crane. Included are the least tern, piping plover, and the lesser prairie chicken, which was recently delisted from the endangered species list!

Why this delisting? It may have something to do with the EPA's views towards oil/gas energy sourcing and wind energy sourcing. There is no reprieve from the Endangered Species Act (ESA). Indeed, far from it. The affected citizens of the Midwest—especially farmers, ranchers, and people employed in the oil, gas, and mining industries—will see their livelihoods jeopardized by the feds' plans. Oil and gas wells will be clustered in groups to avoid scattering them across the sage grouse's habitat. Drilling near breeding areas will be prohibited during mating season, and power lines will be moved away from prime habitat to avoid serving as perches for raptors that prey on the birds. Wind farms will be allowed to expand under the new regulations, of course. Their towers, guy wires, and power lines are exempt. Raptor perches for hunting will continue to be encouraged via the power lines and turbines.

The attempt to homogenize the take process leads one to believe they are seeking wider kill authority than ever. Understand that *take* means "to harass, harm, kill or to attempt to engage in such conduct." This is from the USFWS regulations. These regs are based upon law as written by Congress to protect endangered species, except where wind energy sourcing is concerned.

Most of the central U.S. is part of this region. The state of Nebraska is ground zero for wind development. Nebraska has a Wind and Wildlife Working Group in place to review the manner in which wind farm sites, mitigation, and takes are achieved. The Group works closely with federal and state representatives of environmental agencies and state wide chapters of Nature Conservancy, Sierra Club, Audubon, and the Wildlife Federation. While it does solicit public response, it has no pubic members. It has no authority of enforcement. The Group "supports the development of wind energy resources when the planning and siting process avoids or minimizes impacts to wildlife populations and natural areas."

Their recommendations include wise site choice, an Avian Protection Plan, burying lines, lighting tall turbines, absence of guy wires for support, manage cut-in speeds and blade pitch angles during migration, desk surveys (for whooping cranes, bats, and raptors), intelligent fatality searches, use of evidence of absence software for the mortality surveys, and designation and use of mitigation funds.

This is a recent report, from 8/2015. These approaches are commendable. They are reasonable. These concepts apply far better than simply ignoring the deaths, better than not reporting them at all. They ignore alternative ideas entirely.

They need to ask these questions:

What about alternative sites?
Are scientific studies of value compared to desk surveys?
What about turbine numbers and size in each farm?
Is anyone consideration given to alternative turbine design?
What other avoidance techniques or technologies are important?
Who will conduct fatality searches and how?
What are the alternative energy solutions?
Are there alternatives to death?

Reporting avian mortality is challenging. The wind farms are not required to report, or even investigate, the deaths. Any action they take is voluntary and thus unavailable to the taxpaying public. That would be you, dear reader. You have no right to seek, request, or read any information the industry may or may not have on avian mortality. When and if the industry should partake in such avian mortality surveys, the data is regarded by the current Administration as privileged information. The Administration claims that it does not have the authority to force the release of this data. Isn't that interesting, from a political organization that can force so many other industries to do its bidding? Banks, oil companies, car companies, real estate firms, lending institutions—each are required to file and retain voluminous reports on every aspect of their commerce. Yet wind energy companies live comfortably beneath the green shade umbrella.

In fact, avian deaths are now allowed. They are a *License to Kill*. Reporting on avian mortality is a foolish exercise by the wind industry if only because they would shoot themselves in the foot if they did in fact make such reports. The older reporting is fraught with errors: timing of site examinations, frequency, distance from rotor blades, mortality definitions, area included/excluded, number and placement of surveyed turbines. This is not a scientific approach to a study of any type. It is a series of exceptions that prove their point: no harm here, few deaths, no bodies.

The industry can kill unknown numbers of birds for five years without question. The ITP formulary is simply a prescription for death.

In the next several pages we shall look at the evolution of mortality surveys and of this regulatory take order in greater detail.

California and Wyoming now vie for the lead in annual golden eagle deaths.

Total deaths recorded	1997 - 2012
California	27
Wyoming	29
U.S.	85[140]

These figures bear some ambiguity. They ignore the entire Altamont wind turbine area, as the statistical basis from these killing fields is so uneven. Between 75 and 110 eagles die there each and every year, as an estimate. The local Audubon Society says. "Few would now deny, however, that Altamont Pass is probably the worst site ever chosen for a wind energy according to a 2004 California Energy Commission (CEC) report, as many as 380 burrowing owls (also a state-designated species of special concern), 300 red-tailed hawks, and 333 American kestrels are killed every year. In all, as many as 4,700 birds die annually as a result of the wind turbines."[141]

This report also notes the growth in fatalities with the growth in turbines. The number of known eagle deaths doubled in 2012 from 2011, which saw a similar doubling. Only 19% of discovered bodies or body parts was from deliberate *mortality surveys*. The authors' honesty is to be applauded, given where they work.[142]

Still, avian mortality measuring techniques must be developed and tested, then compared against ancillary evidence. A wide variety of factors must be considered. These techniques are much in dispute across the scientific community. The results are disputed. The effect upon the environment is scaled by the assumed impact of climate change.

The *greater good illusion* is played upon the people. We must destroy to ensure life. What we do is for the greater good of the Earth. Religious overtones abound. Most reporting is either not done at all or kept as a company secret. This behavior is fully supported by the federal agencies entrusted with the protection of the environment and wildlife.

While not an exhaustive list of these considerations in the investigation of avian mortality across the nation, these subjects are *de minima*:

Season (migratory, breeding, off season)
Frequency of site visits
Time of day of site visits
Distance from monopole relative to MW capacity
Area of observance
Type of body part recovery
Scavenger removal and saturation
Weather conditions during site visit

Despite the aridity of detail from the wind industry, there are reports available on avian mortality. Let's examine a few such reports.

The first is a set of three tables from New Jersey. It covers the time frame of January through August of 2009.[143] The examinations were refreshingly scientific.

Date	Time	Turbine	Species/Identify	Notes
03/09/09	1230	T2	Dunlin	Not collected/wings & keel only.
05/27/09	0900	T4	Unidentified small passerine	Not collected/wing & sm. part of body
05/27/09	0925	T1	Unidentified small passerine	Not collected/wing & some feathers
06/08/09	0750	T2	Laughing Gull	Two wings only.
06/24/09	0745	T2	Laughing Gull	Not collected.
07/31/09	1000	T1	Barn Swallow	Not collected. Under truck.
08/18/09	-	T3	Osprey	Collected/removed by ACUA staff.
08/24/09	0000	T4	Green Heron	Not collected.

Table 2

Table 3. Proportion of bird carcasses encountered from August 2007 – August 2009 at each turbine after adjusting for unequal search area (i.e., proportion of area searchable)

Turbine	Carcasses encountered (raw)	Proportion of area searchable*	Carcasses encountered (adjusted)**	Proportion of all carcasses encountered
1	4	0.32	12.37	0.15
2	10	0.31	32.47	0.39
3	14	0.83	16.87	0.20
4	8	0.64	12.52	0.15
5	2	0.24	8.22	0.10

* Relative to total search area (16,900 m2)

**Carcasses encountered (raw)/Proportion of searchable area

Table 3

Table 4. Proportion of bat carcasses encountered from August 2007 – August 2009 at each turbine after adjusting for unequal search area (i.e., proportion of area searchable)

Turbine	Carcasses encountered (raw)	Proportion of area searchable*	Carcasses encountered (adjusted)**	Proportion of all carcasses encountered	
1	6	0.32	18.56	0.16	
2	8	0.31	25.98	0.22	
3	20	0.83	24.10	0.21	
4	20	0.64	31.30	0.27	
5	4	0.24	16.44	0.14	

*Relative to total search area (16,900 m2)

**Carcasses encountered (raw)/Proportion of searchable area

Table 4

This set of tables cover bird and bat mortality. The full text of this detailed report is here.[144] This is a raw data collection discussion, so no conclusions are reached outside of suggestions for fine tuning the collection process and the monitoring itself. The mortality figures are significant, particularly for the raptor population and local bats.

The second report is from the National Renewable Energy Laboratory (NREL). The subject is the nemesis of the wind industry: Altamont Pass. While dated, covering the period from March 1998 through December 2000, it is quite extensive in its recording of avian mortality. The executive summary is clear:

The Altamont Pass Wind Resource Area kills large numbers of birds.

Proximity to turbines aided and abetted bird mortality.

Canyon and rock situated turbines killed raptors.

The clustering of gophers attracted hawks to their death.

Tubular towers were the greatest killers.

Rotor height, wind-swept area, and rotor speed directly correlated to mortality.

Data from 24% of the turbines resulted in 426 carcasses found. The result was an estimated 1,026 deaths each year, 50% of which are raptors. The entire 7,000+ turbine population was not researched. Different birds died from different causes, differing distances, food distinctions, and turbine design differentials. All featured in the killing zone.[145]

This data set is ancient by wind energy standards. It was refreshed in 2006, with newer turbine designs incorporated in the survey. Previous commentary on the absence of a scientific approach have been noted. Since the Altamont surveys, the entire process has gone dark, under the cover of corporate secrets. Promises to open and full reporting from the Administration on government work in 2008-09 are woefully neglected. Birds are the least of the neglected ones, *los olvidados*.

The industry has its own research teams, self-funded and/or state and federally funded. AWEA, most of the wind energy component manufacturers, and many environment NGOs support the American Wind-Wildlife Interactions. This is the NGO for the group. Their most recent public consumption paper is called *Wind*

Turbine Interactions with Wildlife and Their Habitats, dated May 2015.[146] It states: "Individual birds and bats may die as a result of collisions with wind turbines. The potential for biologically significant impacts continues to be a source of concern as populations of many species overlapping with proposed wind energy development are experiencing long-term declines as a result of habitat loss and fragmentation, disease, non-native invasive species, and increased mortality from numerous other anthropogenic activities."

That is as good as it gets. The report then discusses fatalities in terms of number of deaths discovered per MW per year. Recall that MW is the nameplate capacity, rather than the delivered amount. It is a larger denominator. The authors do note this reporting format has difficulties.

You begin to understand their drift when you read that "the number of birds killed at wind energy facilities is a very small fraction of the total, annual human-related bird mortality and two to four orders of magnitude lower than mortality from other factors, including feral and domestic cats, power transmission lines, buildings and windows, and communication towers."

Where have we heard this before? Hmm... Further, fatality rates do not appear likely to lead to species declines (according to these authors). The expansion of wind energy may lead to further discussions on the subject.

To their admirable credit, they do suggest:

Blade feathering and initiation of turbine rotation at higher wind speeds.

Siting away from raptor's ideal locales such as hills may result in lower fatality rates.

Ravens and crows are better at site avoidance than raptors.

Radar and acoustic tools do not appear to be effective in preventing avian-turbine interface.

Alternative turbine design considerations are currently limited to painting the blades ultraviolet.

Bat mortality rates can be significantly higher than bird fatalities, but the kills do not impact survival of any bat species (at least not as much as White-Nose Syndrome). Further, barotraumas and male orientedness do not appear to be significant reasons for bat deaths. The relative high number of migratory bats may be explained by their habit of moving about (season transition).

This May 2015 report reflects "the latest publicly available information about the adverse effects of wind turbines on wildlife in North America and the status of our knowledge regarding how to avoid or minimize these impacts."

That statement is doubtful. It ignores all the data available from other sources. It is inclusive of just two reports from 2014. The remainder are quite ancient. If you read the 2013, 2014, and 2015 reports from the American Wind-Wildlife Interactions, you note the only differences: an inclusion of two reports, one for 2013 and one for 2014. This is an update of all available information?

These are exclusive of any non-U.S. based reports. The determining factor for inclusion is not representative. The data valuation determination is less than persuasive. The comparative statistics, by their own determination, have difficulties.

This data sourcing is less than ideal. We can follow this poor dataset with one from the California Wind Energy Association. Here, they spent a significant amount of money researching the ability of field examiners to find and report tagged, dead birds placed around turbines within the previous 60 days. They used this reporting phenomena as proof or disproof of four different formulas regarding the truthfulness of said reporting relative to assumed results of such reporting. $1.4 million spent on discovery techniques. None spent on actual mortality discovery and reporting.

Who's on first?

CHAPTER TEN
CALIFORNIA CONDORS

https://www.vimeo.com/39678029

Jesse Grantham is a distance swimmer. I am a distance swimmer. We met at the pool. He has done more than any other to advance the cause of condors in the wild and in captivity. For much of his professional career as a field biologist he developed and managed the Condor Project in California. This project has lasted for nearly three decades in its pursuit of rescuing the California condor from extinction. There were just 22 breeding pairs left on the planet when he began his lifelong work. He has persevered to the point of near success. Listen to Jesse:

As of January 31, 2014, total world condor population is 410. Captive population is 178, and wild (California, Arizona, and Baja, Mexico) is 232. That breaks down to 128 in California, 29 in Baja, Mexico, and 75 in Arizona/ Utah. California breaks down to 70 in Southern California, 31 in the Big Sur area, and 27 around Pinnacles National Monument.

I doubt condors could survive 2 to 6 deaths a year from turbine deaths. You couple that with lead poisoning, shooting, and miscellaneous human-caused mortality and you'll put the bird over the top. The problem is condors generally lay only one egg every other year. The chick from that hatching is dependent on the parents for almost a year. A lot for a young condor to learn. And condors don't reach breeding age until they are 6 to 8 years old. So if you kill a few breeding aged adults, it takes a while to replace them, since mortality is not going to be selective. Also killed will be condors from 2 to 8 years old, birds just approaching breeding age.

Condors are the largest raptors in the world. They are also the smallest group of raptors in the world. 100 breeding pairs survive in the wild. With wingspans of nearly 10 feet, they range over hundreds of miles in search of carrion, cleaning the area of waste. At 55+ mph, they move effortlessly, gliding on the updrafts of wind thermals.

Somehow, on May 10, 2013, the USFWS issued permits for incidental takes to the wind farms in Tehachapi and Tejon Passes in California. USFWS Director Daniel Ashe said, "This is the first time we've authorized incidental takes of California condors." The agency then proceeded to invite other farms to apply for the same permission. Three decades of condor preservation work by state and federal agencies and personnel has brought the species from the point of extinction—22 breeding pairs—to the point where the USFWS feels comfortable killing the giants once again. Daniel Barnett of the Kern County Audubon Society offered, "I can't believe the federal government is putting so much money into an historic and costly effort to establish a stable population of condors and at the same time issuing permits to kill. Ludicrous."

Ashe, of the USFWS was encouraged by the promise of radar and experimental telemetry to track the birds and the concurrent promise to shut down the turbines when the birds approach. Said systems are built with the hope of identifying the birds soon enough to shut down the turbines spinning at 170 mph while still minimizing costs and maximizing profits by turning the giant blades back on as soon as the avian troupe leaves.

The experimental telemetry is dependent upon collars worn by a few individuals in the condor family. These collars need to stay on and remain functioning for the signals to be intercepted. Replacing a collar is tedious, time consuming work, often subject to failure. Condors are not keen on human companionship, despite most of them having been raised in captivity (perhaps this is the reason). Collar wearers are not necessarily representative of the population. Those killed without collars are unreported.

As with their European and African cousins, large raptors are herd members. They seek food in kettles, formed in the sky, on the fly, as soon as one bird sights a meal. They swoop down and eat en masse. A turbine killing one condor has a remarkably high chance of killing others as they rush to the scene of the crime, only to be lacerated by the same blades of death.

How insane is this permit? What rational thought went into this scheme? Are we so blinded by our religious devotion to global climate change that we must kill birds to save them? Is the ghost of Jim Jones running this program?

This is an abrupt about-face for USFWS, whose representatives stated in 2012 that issuing *lethal take permits* for the California condor to wind developers or anyone else was out of the question. The agency had warned Ventura County just a few months prior that careless approval of a 350 megawatt wind project could land the county in hot water with regard to illegal take of condors, adding "We cannot envision a situation where we would permit the lethal take of California condors."

Director Ashe used as a prop at the announcement meeting a captive condor from the local zoo.[147] His grandchildren may be surprised to find this captive the last remaining member of a species destroyed by regulatory diktat in just a few years. They will question the ethics of such destruction, all in the name of a godless religion, a belief in the Evil of Carbon.

In all fairness, the *New York Times* has presented another point of view on the topic of condor takes in the Mojave. It characterized local environmental activists' commentary over the potential loss of a condor: "The distressed reaction was familiar: wind-energy opponents often cite avian mortality to bolster their case. But the origin of the Interior Department's decision to grant the take permit near the Mojave Desert remained unclear, since no condors have died there."

Apparently reading the BiOp (USFWS's Biological Opinion) was unnecessary: "The Service anticipates that over the 30-year life of the project, one California condor is likely to be killed as a result of being struck by a turbine blade. If this single bird has an egg or young nestling at the time of its death, this egg or young nestling may also die if the California Condor Recovery Program is unable to recover it."[148]

A visit to the Mojave was not in the reporter's brief. Comments on the commonality of turbines in the area since 1989 and comparisons to its size relative to Manhattan Island are parochial, at best. The turbine numbers have significantly expanded in the past six years, as has the condor flyway, both intersecting over a range far greater than her tiny metropolitan isle. This time lapse[149] from 1984-2012 shows the rapid increase in support roads for, you guessed it, wind turbines. You can view a story from so many angles.

An even newer report indicates "deaths/gigawatt-hour of power generation declined quickly with increasing capacity factor among wind turbines, indicating collision hazard increased with greater intermittency in turbine operations."[150] Greater fluctuations in turbine power supply due to wind strength variability increase bird deaths. We have seen that capacity factor is far more indicative than rated capacity of both energy supply and avian mortality. The inverse relationship is damning. More deaths result from intermittent power supply. The authors note that their observed results for raptor deaths were three times greater than they expected. The study encompassed just 21.9 MW of turbine tower rated capacity in the Alta Mont wind farm region, a very small scale indeed.

Chris Clarke, the local desert expert on alternative energy projects and their impact on the fragile ecosystem of the deep desert, says, "A more likely broad cause of the USFWS reversal on condors is political pressure to develop renewable energy generating capacity at all costs. Fish and Wildlife Director Daniel Ashe said the decision reflects a difficult reality. The threat of prosecution jeopardized the construction of large-scale alternative energy facilities in the wild and windy places preferred by condors.[151]

Yes, that is the exact point of the law: the *threat of prosecution* to protect the local environment and a varyingly fragile ecosystem. Avoiding the issue by authorizing takes (avian deaths) breaks the law and undermines its very purpose. How can we preserve the planet for our children if the law is so effortlessly sideswiped by the very government that writes the regulations?

This chapter has been devoted to condors as a warning on the intensity of the religious fervor devoted to the destruction of birds in the interests of political, religious, and pseudo-scientific agendas.

CHAPTER ELEVEN
BATS

Myself, I rather like the bat.
It's not a mouse, it's not a rat.
It has no feathers, yet has wings.
It's quite inaudible when it sings.
— **Ogden Nash**

Bats are the only flying mammals. They give birth to one pup each year. The pup weights as much as 25% of the mother's weight. This is like a human mother bearing a 30 pound baby! The pup is suckled by the mother for as long as six months in maternal broods.

Even at the small bird scale, bats suffer the greatest predation of all avian species in the turgid grip of the turbine. More than 1.7 million bats have died in turbines prior to 2011according to Bat Conservation International. 1.3 million bats died in 2012. Figures for later are unavailable at the site. 23 of 47 species are killed by turbines, nine of which are endangered species, on the federal endangered species list, or both.

Their small brood reproductive rates compare in number and annual frequency to raptors reproduction, although their colony sizes far exceed raptor populations. A single birth is typical each year. High adult survival rates replace high reproductivity for species survival.

Why the concern about bats? They occupy an important niche in the entire food cycle, consuming insects that prey on edible plant life such as corn, wheat, oats, maize, and barley. They are the small scale version of vultures, buzzards, and condors. They wash the fields of grain devouring insects. Without bats you have to increase the use of pesticides several orders of magnitude. $3.7 billion is saved each year as bats take the place of pesticides. Call them the GMOs of the avian world. They are not well liked but they are critical to survival of the food sources of the species *homo sapiens*. Virtually all bats each insects or fruit.

How do bats die? They have echo sounders and can rapidly negotiate the fly-bys of turbine fields. They often fly up to the turbine tower seeking rest. They die from barotraumas. Their lungs explode. The differing wind pressure between the forward and trailing edges of the turbine blades explodes their lungs. In short, the down sweep of a blade cause their lungs to explode.

Laura Ellison writes for the USGS bibliographic summary of all bat/turbine literature: "Since the beginning of large scale wind-energy production in the 1980s, biologists have expressed growing concern over the potential impacts of turbines on bats. The evidence suggests that bats of certain species are dying by the thousands at turbines across North America. Migratory species that rely upon nightly roosts are most affected. Turbine-related bat mortalities now impact 25% of North American bat species. Most documented bat mortality at wind-energy facilities has occurred in late summer and early fall."

Let's review a few of the more recent scientific bat studies. Our first report is on the economics of bats. The study in the April, 2011 edition of *Science*[152] illustrates the economic disadvantage of losing bats. Because of the combined effects of white-nose syndrome (WNS) and wind turbine mortality, sudden and simultaneous population declines are being witnessed on a scale rivaled by few recorded events affecting mammals.

One million bats have died from WNS since 2006 and winter colonies of local bat populations have declined by as much as 70%. The disease has taken a serious toll. Wind turbine mortalities have been previously stated as 1.3 million in 2012. Deaths of migratory bats are expected to range between 33,000 and 110,000 annually in 2020, simply in the Mid-Atlantic Highlands.

So what. They're bats. Who cares besides a few biologist geeks? You need to care. A bat eats 6 grams of insects each evening meal. 150 bats consume 1.4 million insects over an Indiana field overnight. They can consume half their body weight each night. That would be the same field that produces your corn flakes and oat granola. The *Science* authors give a median cost of $74/acre. They allow a range of impact from $3.7 billion to $53 billion each year from the loss of bats. You can half or quarter these figures to account for WNS. You still arrive at an enormous amount of agricultural damage from bat deaths from wind turbines.

This damage is what John Droz was referring to in his earlier discussions on the economic impact upon three North Carolina counties from wind turbine placements. This damage is real. It is growing as the number of turbines grows. Extinction is a very real possibility when death from both sources is compiled.

Bats take the place of pesticides. They represent the UN's best guidance on biodiversity. Turbines are destroying biodiversity in the fields of America. WNS is destroying biodiversity in American fields. The latter can be biologically adjusted to via survival rates. Those who survive develop immunities to WNS. Can the same be said for turbine interactions?

A simple solution comes from our European allies in the Carbon Wars. Bats fly at certain wind speeds because they are very small flying mammals with very little mass. They love to fly at wind speeds below 6 kps (kilometers per second). Above this speed and they stay put, happy to hang around with their inverted friends. Start the wind turbine blades at this speed. 97% of bat fatalities are avoided in Spain with this simple technique.

In our terms, this means increasing the cut-in wind speed from 9 mph to 11 mph. According to Edward Arnett of Bat Conservation International, the results parallel those from Europe: 68% to 97% decline in bat mortality with this simple solution. Add an awareness of migratory periods for the little tropical touristas and you can effect a significant reduction in bat mortalities.

Germany has seen an increase in turbines from 1,200 to 23,000+ since 1992. The effect upon bat populations, particularly migratory groups that move gregariously from the tropics to the north, is egregious and growing. The ability to determine the impact is more difficult than simply counting carcasses.[153] A variety of counting techniques are employed in Europe.

There is more to bat mortality than meets the eye. A more recent report from the scientific literature in Europe points to mortality affecting both local and noctule bat populations, with greater impact to female migratory species and juvenile local species.[154] The authors urge efficient and effective mitigation measures upon mortality of both species. They lament the inexactitude of current studies. They regret the inability of authorities to approach, much less handle the problem. This lamentation is universal among biological researchers for birds and bats.

In the U.S., the National Academy of Science published a report in 2014 on the bat mortality issue.[155] The 12 authors did a meta-analysis of all studies to interpret the co-variance of bat behavior and turbine speeds. Bats have evolved to sink downstream of an eddy for feeding and mating rituals. This is a death spiral in the vicinity of turbines. Their sweep and speed crush a bat's ability to detect blade distance. The authors concurred that bat incidences occur at lower wind speeds and turbine rotation initiation.

A study from 2010[156] of bat mortality from wind turbines in Pennsylvania demonstrated a striking death rate of at least one bat per turbine each night of the study. They found that curtailing cut-in speed to +6 meters per second reduced mortality by 72%. This is in line with the Spanish study. Some wind energy firms are taking this into consideration, allowing for the reduced economic benefit of higher cut-in speeds. The reduced earnings are less than 1% according to ABC.[157] Many find there is little revenue lost in the six meters per second rate. They at least understand the concept of trade-offs: a kinder environmental touch results in a benefit to both human and bat societies.

Deterrent devices are emerging as the industry tries to deal with the mortality problem in its own convoluted manner. A study from BWEC of 2011[158] illustrates the issue. Ten turbines were rigged with ultrasonic devices to deter bats from approaching. Carcass count was performed for several weeks in the fall of 2009 and again in 2010. The devices show a range of bat responses, from 2% to 62% reduction in mortality, but adjusted for effectiveness, the figures were meaningless to the researchers. Humidity, weather, and distance from blade sweep were of greater impact (or reduced the impact of the devices). Devices broke and were not replaced.

A concurrent research occurred in Indiana.[159] This covered a larger area and number of turbines: 355 units. To their credit, the researchers performed weekly and daily carcass searches. Most simply use a formula to cover for infrequent or non-existent searches. There are a least four of these formulae and more debate about their use than their substitution for real time search efforts (too costly and ineffectual!). Their findings are shocking. During the spring and fall (180 days) of 2010 between 18 and 25 bat deaths (median = 21) per turbine per night were recorded. Low wind speed and higher humidity brought out the bats. Increasing cut-in speeds reduced mortality rates by as much as 90%.

Let's do the math:

355 turbines x 21 deaths/night = 7,455 daily deaths
7,455 x 180 days = 1,341,900 migratory deaths

The authors conclude that Indiana bats are not at risk from turbine mortality. 16 raptor and passerine carcasses were found, but were ignored for the bat study. Half of the bat carcasses were found outside of the temporal parameters of the search and were ignored. As the study was designed to study just one species, it reported no significant impact to local bat populations. Daily searches resulted in far greater carcass counts than weekly or computer models from formulae.

Let's examine the concept of *search*. This illustration shows ways in which searches can be performed. *Squaring the circle* ignores significant land area. Large birds are flung further from the blade impact point than smaller birds. They can also attempt to fly or crawl away injured. The circle is also reduced in size, which ignores the fling arc as well. More than a third of the coverage area is ignored by this simple sampling technique.

Using Square Search Plots to Produce Data
from Undersized Search Areas

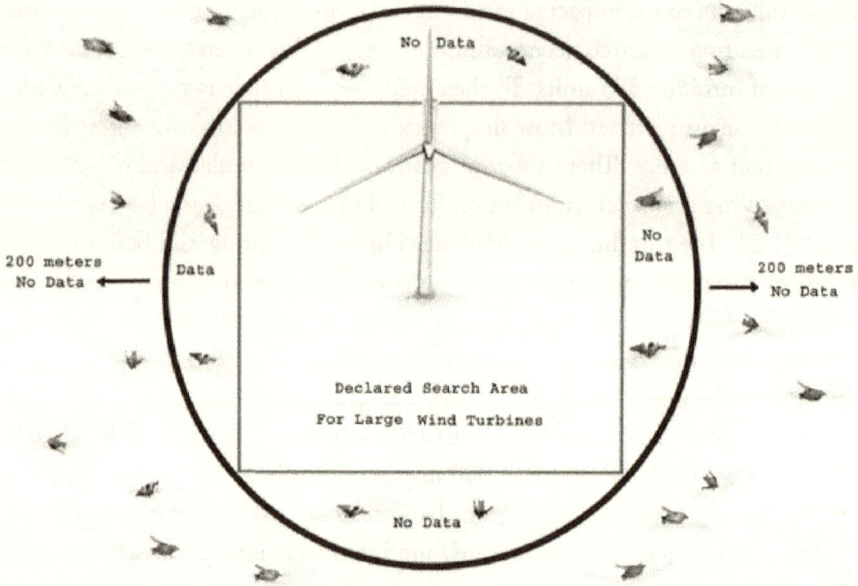

One wind industry trick is to declare square search plots and then
give figures for circular search areas. They then claim to have
searched out to the maximum distance. This eliminates 36% of the area
out to this distance. An area where most of the carcasses will land.
Square plots can also be rotated to avoid direction of carcasses throw.

Thanks to Jim Wiegand for the illustration.

Using one of the four agreed upon formulae for carcass discovery eliminates
this needless walking and counting waste of time almost entirely. One only has
to examine the square patch visually once a week or once a month, report back
the findings, run the computer model, and your research is complete. On to the
next killing field.

The nature of scientific studies is to randomize actions, narrow focus on the
results, and report observations. These authors did just as expected. Yet upon reading
their work, you may be overwhelmed by the missing data, the lack of using all data,

and the narrow range of reported results. Perhaps it's just you. Or, just maybe, the glaring data breaches are shouting out to be broadcast.

More than one million annual bat deaths can be inferred. Nearly three thousand annual avian deaths can be inferred. Each inference is based entirely upon their data. The annual death counts from ABI verify these estimates: 1.3 million dead bats in 2012 in just one locale in the Hoosier State. A quick glance at the USGS interactive wind turbine map of the U.S. (http://eerscmap.usgs.gov/windfarm) reveals three massive areas of turbine development. Most share easy road access. You can see for yourself the layout across the corn and soy fields. Are weekly carcass counts and computer modeling the only way to understand mortality?

A further study of small wind turbines (SWTs) is the first of its kind. The authors find that smaller units have similar bird and bat mortalities as larger units, that bats fly in close proximity to blades at lower speeds, and that birds have avoidance skills according to species. They also lament the complete lack of planning guidance from authorities.[160]

Let's give the wind industry its say on the matter of bats and avian mortality. The two bat studies referenced on the previous three pages are from AWEA, AWWI, and BWEC, with financial support from the industry and USFWS. In addition to the AWEA and AWWI, the USFWS also supports the Bat and Wind Energy Cooperative (BWEC). Such a community sounding title. BWEC acts in association with yet another non-profit, the National Wind Coordinating Collaborative (NWCC). The latter's title begs a distinction between the more mundane soviets of central Asia. Each of these groups holds a biennial meeting on the subject of their interest, bats and turbines.

These meetings are flush with speeches and presentations. In full disclosure, I have participated in such meetings on other subjects and we do tend to spend some time preaching to the choir and drinking with the preacher. Take the following with a grain of salt. Nevertheless, the most recent meeting in Colorado in 2014 had plenary sessions on birds, bats, and wind. The *Proceedings* were recently published, six months after the meeting. The bibliography from the meeting is resolute, if dated. It includes more than 100 articles published and unpublished from the white and grey world (the distinction has to do with the source of funds).

There are over 100 sources, but they are antiquated. Two each date from 2013 and 2012. Most are from much earlier and several from the last century. Perhaps this is because the syllabus is updated from 2004. This is an immense period of time for a rapidly evolving industry. Quoting a line from a 1992 study is facetious at best, fallacious at worst. The commentary revolves around how to measure deaths, how to count them, and how to use equations to take the place of actually being in the killing fields.[161]

The abstract begins with the observation that wind energy contributes to the long term viability of every species on earth. Wind also reduces mercury and hydrocarbon emissions while improving water use. "Impacts to wildlife populations have not been documented." "The potential for such impacts continue to be a concern." These concerns "must be balanced against the many other anthropogenic negative impacts" upon avian and bat life. "The proceedings document the current research in understanding risk, fatality estimation, population demographics, detection technologies, and impact minimization."

Among the significant reports visited in the *Proceedings*, the following comments are representative:

"Sustainable mortality varies between species."

"Raising cut-in speeds results in economic loss."

"Effective monitoring likely oversamples for multiple species."

"Operational staff are more cost effective at site monitoring than paid consultants."

"Allowing golden eagle takes of 5% of the (national) population still provide for stable breeding pairs."

"Power line density could impact eagle mortalities."

"It is possible to make site-specific recommendations for reducing avian mortality."

"The best scientific information may be insufficient to isolate wind farms as the cause of territory abandonment."

"The USFWS Bayesian model is a best guess of anticipated mortalities."

"Developers may be requested to redesign sites."

"We are likely underestimating the number of dead raptors."

"Conservationists models overestimate risk to birds."

These conclusions and comments are interesting. They show an awareness of the avian mortality problem not expressed in the service or industry literature. This is a mature view. They indicate a lack of serious data. Many of these reports are meta-analyses or reviews of older reports in conjunction with each other (much like we are doing here). They demonstrate varying degrees of industry prejudices, from profound to specious. They show a willingness by some to do real research on avian mortality and a corresponding lack of initiative by others.

These conclusions differ in degree and temperament from the European commentary. They are far less scientific, although a few are double blind and peer reviewed. No one ranted against their hosts or the government. One report from an NGO made much of species destruction from climate change while urging more wind. They should change their name or their tune or their venue. Death is Good. This hardly makes the front page headlines, yet that is what this non-profit group promoted, as implied from its presentation.

The nature of a few of the discussions was reasonably serious. Nevertheless, the choirs and preachers were fully aware of their audience and vice versa. The public is entirely unaware of this biannual event. A shame about this. The dearth of good data was clear from the actual remarks of the scientists in attendance. Were this information more widely transmitted, the industry would have far fewer issues with public perception.

The facts remain clear. Little is known, or at least shared with the public, at the industry level of the effects of turbines on avian mortality, much less of deterrence. All suspect it is greater than stated by the Service. Access to this forum is as restricted as it is for all information sources from within a hidebound bureaucracy and tight-lipped industry. They have only themselves to blame for the miserly perception by their public audience and the miserable perception they instill in an untrusting scientific public.

In closing this bat story, enjoy this final note: http://batslive.pwnet.org. This site is intended for our youth. The learning videos are freely available to schools and churches. They tell friendly stories from Bat World!

The next scientifically important environmental research study was released in November, 2013 on the subject of avian predation by wind turbines, specifically monopole towers. Let's review it in some detail. This work by our previously cited scientific associates, Loss, Marra and Will, is worthy of serious consideration.

They have performed the most thorough and complete scientific study of avian mortality to date.

The Smithsonian Report, 11/2013

Scott Loss and Peter Marra of the Smithsonian Conservation Biology Institute's Migratory Bird Center, working with USFWS biologist Tom Will, recently reported their study of "Bird collision mortality at wind facilities in the contiguous United States" in the journal, *Biological Conservation.*[162] (See also Appendix A.) This peer reviewed study is the most recent and most extensive work to date on the avian/turbine interface. Its conclusions disagree with virtually every statement from the *WEIR* (wind energy industry/regulatory) perspective.

The *Biological Conservation* study finds that a *mean* of 234,000 birds are killed annually by wind turbines in the U.S. It states clearly that

Avian mortality has increased with monopole wind tower construction.

Tower height increases mortality.

Great Plains deaths are lower than those in mountainous regions.

Species risk evaluation is warranted with the widespread shift to taller, monopole rigs.

As lattice towers are replaced with monopole structures and their heights are simultaneously increased to take advantage of stronger, more consistent wind patterns, bird deaths continue to increase. Previous studies did not differentiate between the two structure types. Monopoles are the vast majority of turbine tower support today. The *lattice as nesting ground* theory is disabused by factual evidence. The height solution is disavowed, as well.

Current pre-construction collision risk assessment is unreliable, as it avoids *species risk* by focusing upon *general population risk*. It also relies upon outdated methodology or upon mere estimates. As an example, the USFWS estimate for total bird mortality is 573,000, but this estimate is faulty. 68 studies of bird/tower collisions were researched. Previous studies on lattice work alone—such as at Altamont, CA—were ignored. This may have significantly impacted avian mortality count.

They note that as more months in the year were sampled, more deaths were recorded – and as fewer months were sampled, fewer deaths were recorded. This bias reflects previous underestimates of bird deaths. The manner in which deaths were discovered (directly or indirectly, purposeful or incidental searching) and reported varies between the meta data sources. The type of searches, their radius, frequency and manner of reporting also varied. Scavenger removal rates vary. The types of mortalities and or injuries also varied.

How this impacts future studies remains unclear. Extrapolation of information, its representative nature and its interpretation will always be improvable. Transparency of industry data collected by company representatives is wholly absent today. Regulatory requirements for such transparency are wholly absent, as well. Better data, more meaningfully collected, and purposefully analyzed across objective standards should be valuable to future researchers.

Their ultimate database included 1,000 onshore wind facilities with 44,577 turbines. Thus, they analyzed 38% of all wind farms, but nearly 90% of all turbines. 3,605 bird deaths were recorded, from 218 species. A brief summary of the data is presented here:

Estimates of bird mortality from collisions with monopole wind turbines in the contiguous U.S.

Region	Total # of turbines	Total MW capacity	Total mortality	Mortality per turbine Mean	Mortality per MW Mean
California	13,851	5,796	108,715	7.85	18.76
East	6,418	11,390	44,006	6.86	3.86
West	5,757	9,590	27,177	4.72	2.83
Great Plains	18,551	29,896	54,115	2.92	1.81
Total U.S.	44,577	56,852	234,012	5.25	4.12

While the height of turbines appeared to correlate to avian deaths, the size of rotor diameter may also be a cause. Of course, both are correlated to each other, so the point is a small one. Deaths in the Midwest were significantly lower than in the Rockies, perhaps because of migratory patterns, cliff ledges for nests, relative abundance of food, etc. The smaller death figures may still mask the greater potential for species wide impact.

Suggestions from the authors include:

> Pre-construction site assessment one year prior
> Ongoing analysis during operations
> Species level impact reporting; raptor impact in particular
> Bat impact studies need to be improved
> How, why, and results to insectivore populations
> Avian mortality is not non-trivial
> 1.4 million deaths from 20% wind power sourcing by 2030

This is the most recent and most exhaustive study of avian mortality from wind turbine towers. While detailed, the authors' observations and their results are shocking. They are inconsistent with all comments from the WEIR.

Promises have been broken, yet again.

As if this were not enough, the solar industry is now participating in bird killings with their newest and biggest Ivanpah solar plant in the desert between Los Angeles and Las Vegas. Freeway drivers between the two cities have watched for years as these three 360 foot towers have risen from the sands. This newest in solar technology is called *Concentrated Solar Power* (CSP).[163] The towers are surrounded by a vast array of mirrors, giving the appearance of an alien landing site. These mirrors reflect sunlight to the top of the towers. Computers aim the *heliostats* at the towers, where the concentrated solar flux heats steam that turns turbines, producing electricity in the conventional manner. That flux has proved a significant environmental concern, as birds and other wildlife have fallen victim to injuries from the concentrated solar radiation.

On February 12, 2014, U.S. Energy Secretary Ernest Moniz ushered the 377-megawatt solar power plant into its next phase, as he formally opened the plant for business. "The Ivanpah project is a shining example of how America is becoming a world leader in solar energy," said Moniz. "This project shows that building a clean energy economy creates jobs, curbs greenhouse gas emissions, and fosters American innovation."

The Department of Energy backed Ivanpah's construction with $1.6 billion in loan guarantees from its Loan Programs Office. At the ceremony, NRG Energy president Tom Doyle credited that loan guarantee with the plant's successful completion. Though BrightSource developed the tech used at ISEGS, the plant is co-owned and co-managed by NRG Energy with Google owning part of the plant as well.[164]

Ivanpah is now responsible for four dozen avian deaths just during its three month commissioning process. Burned feathers and wings have been found along with 38 carcasses of peregrine falcons, hawks, nighthawks, grebes, warblers, and sparrows. A two-year study has been initiated to "examine the solar plant's role in avian mortality." Must have been alien space rays that took down these creatures of the night sky.[165]

The federal Bureau of Land Management (BLM) held a recent meeting in Palm Desert on the next in a series of solar plants for the desert. The proposed 500 MW facility, the Palen Solar Electric Generating System, will add another illusory mirrored lake for birds. The birds see the reflection, fly towards it, and are fried by the thousands in the superannuated heat.

The meeting was held to solicit public comment on the project. No transcript of comments was made for the formal record unless the speaker had brought a written version. This quirk in public hearings format appears odd. According to the National Environmental Policy Act, final documents for such proposals must include public comments only if in the written record. Potential law suits are also allowed only upon formal public written transcripts of such meetings.[166] Absent such written memoranda, what recourse does the public have against its own government?

This story was preceded by the news in July of more than three dozen water birds found dead at the newest solar projects in the Southern California desert, the 550 MW Desert Sunlight Solar Farm and the 250 MW Genesis Solar Energy Project. These deaths have occurred in just three months.

The water birds killed and injured range in species from yellow-headed blackbirds to the once-critically endangered brown pelican. Other water birds found dead or injured at the two projects include eared, western, and pied-billed grebes, the duck species surf scoter, red breasted merganser and bufflehead, the black-crowned night heron, double-crested cormorants, American coots, and the federally endangered Yuma clapper rail. Other birds reported dead or injured at the two facilities in that time period include warblers, goldfinches, a common raven, and a barn owl.

In addition, representatives from the solar array farms told of a few incidents not yet included in compliance reports, including deaths of three juvenile brown pelicans and a black-crowned night heron at Desert Sunlight, and another brown pelican found July 10th at the Genesis project.

Most of the mortalities were discovered by project biologists or other staff, and consisted of finding carcasses in varying stages of decay. At least one bird, the red-breasted merganser found in April at Desert Sunlight, was alive when discovered but died shortly after.

The deaths are attributed to the illusion of water that the solar arrays give. The fact that virtually all of the deaths are water fowl gives substance to the theory. Since this report was made just six months ago on the local public broadcasting channel, over 100 more fowl have died at these two solar plants.[167]

Will this new solar threat to passerines and raptors of the desert create even more ecocatastrophe for California? Fortunately, on December 13, 2013, the agency denied the permit to move forward on the Palen Project, given the mounting evidence of bird deaths from nearby solar facilities including the newest Ivanpah units.

Perhaps Pogo was an optimist.

CHAPTER TWELVE
FALCONS

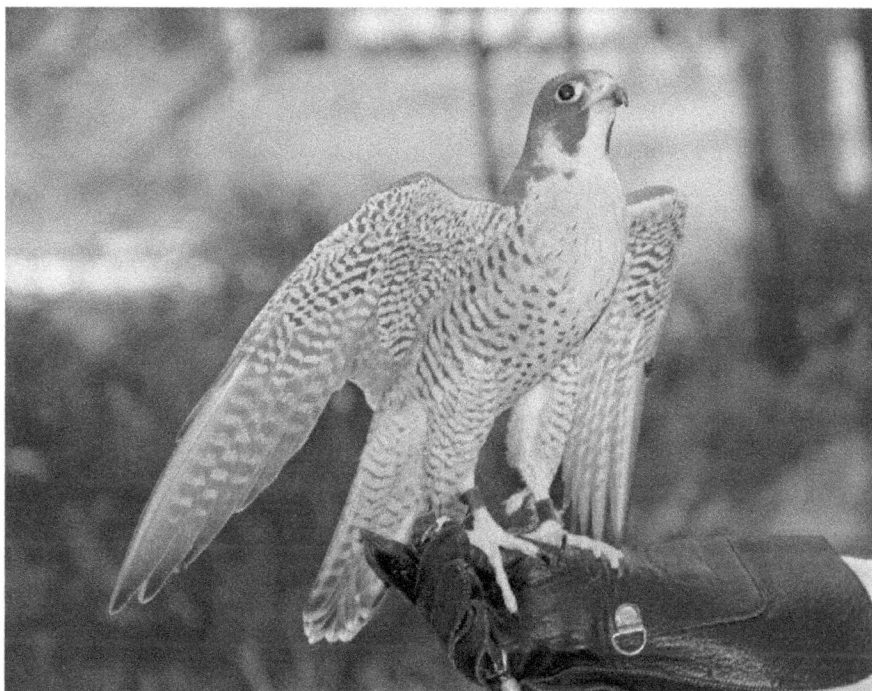

Courtesy of the Carolina Raptor Center and Dick Daniels.

Detective Tom Polhaus picks up the falcon: Heavy. What is it?
Sam Spade: The stuff that dreams are made of.
Detective Polhaus: Huh?

— *The Maltese Falcon*, John Huston

Turning and turning in the widening gyre

The falcon cannot hear the falconer;
Things fall apart, the center cannot hold;
Mere anarchy is loosed upon the world…
The best lack all conviction, while the worst
Are full of passionate intensity.

— William Butler Yeats, *The Second Coming*

Falcons have bred humans for millennia. They were used as secure food sourcing and mating safety. The human preserves and protects these fastest and wariest of the raptor family. Myth and legend intertwine. Nobles and peasants raise and fly falcons across virtually all cultures. Desert, highlands, field, and stream all provide sustenance for the breed. In Persian, Shaheen is the emperor of the skies.

The peregrine falcon is certainly the most successful bird of Mother Earth. *Peregrinus* is Latin for wanderer. Except Iceland and Antarctica, falcons are ubiquitous. The species comes in a myriad of forms. Caracara in the Central American jungle. Barbary falcon from North Africa. Shaheen in Iran and Afghanistan. The Arctic gyrfalcon of its name site. Sakers of the Bedouin (from *saqur*, Arabic). The North American prairie falcon is well known as the malcontent of the falco species. Australasian falcons are known as black falcons. The mysterious South American orange falcon hides in the three-tiered Amazon forest. The small kestrel roams Europe.

Falcons, like hawks, experience the world at a far faster rate than we humans. The density of cones and rods in their retina greatly exceeds our own: one million vs. 30,000. They see in ultraviolet and polarized light. They see stereo-optically. They constantly turn their heads to parallax their prey. We see in a three-color matrix of red, green, and blue; they see in a fourth, ultraviolet. They also see humidity and sense the magnetosphere for navigation. The beak can sever a spinal cord. Their feet are designed to their prey and killing locale: air, ground, sand, or snow. Once engaged in the flesh of a victim, the claws hold as tightly as a vise, without muscular exertion. The wing bones are connected directly to their lungs, allowing the falcon to breath while flying at speeds of over 160 mph. They breathe without exhalation. Their urine etches metal.

A falcon's flight profile is anhedral, or "^" shaped, designed for flapping and speed gliding. They dive upon prey from behind and above. The falcon's hunting characteristics are codified in human pilots' flight manuals. Attack from the sun. Approach from the blind spot just behind and below the body. Attack hard with claws leading to kill instantly. Accelerometers attached to falcons go off scale at 26 Gs. Humans pass out above 6 Gs.

Falcon *contour* feathers differ from most other raptors. They are tight, molding the bird in flight to a low resistance airfoil. *Down* feathers provide insulation beneath their *flight* feathers. *Filoplumes* sense air pressure. Falcons groom endlessly, using their uropygial gland's fatty acids for waterproofing and sunscreen. The dark malar coloring beneath their eyes shades from glare, just as dark under-eye coloring is used by professional athletes. The falcon's back may be shaded blue to blend into the surrounding sky.[168]

Falcons migrate with their meals. In the Tien Shan and Rocky Mountains, they move with the snowline. Greenland falcons will be found in Afghanistan and South Africa. The falcon will cover hundreds of miles each day, crossing seas and oceans with strong wind currents. This is their greatest strength, and their greatest weakness, as we shall see.

Thanks to NatGeo for this amazing series:

https://www.youtube.com/watch?v=e-x-vROfwes.

Regulatory Behavior

The wind industry has been given an exemption from federal laws on wildlife protection. The Eagle Protection Act and the Migratory Bird Treaty Act have been ignored. The word is *take*. Raptor kills are allowed. Bat kills are ignored. Small bird kills are forgotten. Condor kills are now allowed! The reason for these kills is simple: Wing tip speeds in excess of 170 mph at the outer edge of the blades. The speed alone will kill a bird. The vortex created will suck the small songbird in; as the bird loses attitude control, the following blade shreds it. Painful, slow death is nearly always the result.

Under both the Migratory Bird Treaty Act and the Bald and Golden Eagle Protection Act, the death of a single covered bird without a permit is illegal. This was written into law because of significant declining numbers for the genus *raptor* and the species, bald eagles and golden eagles. Their maturity takes five years. They lay but one egg every other season. Their migration takes them directly through the largest wind farm on the planet—Altamont, California. There, they are slaughtered. More die each year. They have no defense. Their greatest enemy is *Homo Environs*.

Under the new guidelines, wind-energy companies—and only wind-energy companies—are held to other standards. Their facilities don't face additional scrutiny until they have a *significant adverse impact* on wildlife or habitat. Under both bird protection laws, any impact must be addressed. One death resulting from manmade activity is to be reported and severely punished—to protect the species. Programmatic deaths are an acceptable cost for renewable energy. Systemic death is permitted with results across the top and bottom of the avian food chain. Raptors, passerines, and bats die.

Imagine if the global fishing industry was given carte blanche to kill any whales and plankton; we must kill them to save them. Imagine if the African savannah were the killing ground for all tusked animals: elephants, rhinos, et. al. Imagine if both polar bears and seals were slaughtered by the thousands.

What would you do?

The Mortised Falcon with apologies to Bogart and Hawk

The Administration overruled its regional and state experts at the USFWS. The wind energy industry was part of the committee that drafted and edited the guidelines. It got everything it wanted. "Clearly, there was a bias to wind energy," said Rob Manes, of The Nature Conservancy who served on the committee.

The story becomes more interesting each time it is told. Perhaps that's because it is told so infrequently. Patched together from press releases and smuggled internal memos and expurgated Freedom of Information Act pages, this is a tale to make Sam Spade nervous! Let's call it *The Mortised Falcon*. The resulting regulation was mortised together from shards. It was presented to the public as a set piece. The hand was dealt on a Friday afternoon at 4 pm to an empty press room. Meetings had been held. Agendas were put forth. Studies were made. Experts consulted.

Promises were made—and broken.

The Freedom of Information Act was devised by Congress to attempt to ensure that the American public has access to information shared with and within the federal government. As long as this information is not viewed as a security issue, it can be seen through a petition process.

The American Bird Conservancy, the hero of our tale, underwent this process recently. They requested all documents leading up to the FWS decree on avian mortality on December 13, 2013. The results were released to it and made public in late January, 2014. It is to be found in the endnotes of this book.[169] You are dared to read them in their entirety of 829 pages. Grab a cup of coffee as you dive into the morass. Your response is invited. Wear your shit boots. It is deep. Much of it is redacted, i.e., blacked out.

No matter your views on the environment, on science, on wildlife, on any of the non-profits, federal agencies, industries, or companies involved, you should be shocked and outraged. This commentary is about none of these groups. It is about the raw abuse of power. It is about federal agencies doing the bidding of an industry over which it is designed to regulate. The fox rules the chicken coop.

The environmental community was handled. They were the stooges for the agendas of the wind industry and regulators. The Wind Energy Industry and Regulators (WEIR) ensnared even their cousins, the environmental NGOs, as shills. You can read their outrage as they see how they are being mistreated. Shame about their lack of outrageous follow through.

The birds, too, are entrapped in these snares, these weirs. Weirs are designed for death. At this industry's level, mortality is righteously abused. It is justified by garbled words of legalese. These hundreds of thousands of raptors and peregrines will become tomorrow's *rara avis*. Raptors will go the way of carrier pigeons, as just another species in the way of industrial power. Let the hunting begin.

The agencies are acting at the bequest of the industry. The industry is suggesting regulatory guidelines for itself to follow. The members of the industry write their own rules, in secret. Public participation has been shut down. Congressional oversight has been delegated to the Energy Secretary's legal representative. Administrative justice prevails through the silent presence of the DOJ. Gotta make sure the law is followed, to the letter.

Let's view a kangaroo court in session. Let's be the fly on the wall at the Star Chamber. Let's see raw power in action.

Meetings, Agendas, and Promises

Since 2009, the Service (USFWS) has been struggling with an impossible challenge. It must referee the deadly duel between wildlife conservation and renewable energy development. Enhanced wind turbine development is a direct result of the Administration's rush to alternative fuel sourcing. Wind, solar, geothermal, ethanol, and other energy sources are being hurried into the national power grid. The clearly stated reason is climate change: "The USFWS supports the development of renewable energy resources as a means to reduce carbon emissions and their impacts on the landscape."

This is not without challenge for the Service. The Bald and Golden Eagle Protection Act (BGEPA) mandates the preservation of these bird species. Wind farms kill significant numbers of this population. They will move towards extinction at mankind's kindless hand. "Methods of avoidance and mitigation of eagle fatalities" is the Service directive.

This impossible duality forces the Service to square the circle. In 2009 it promulgated the Eagle Take Rule (ETR). This changed the permit rules regarding deaths of bald and golden eagle. It permitted takes where the taking is associated with but not the purpose of an activity. Programmatic recurring takes are now authorized that are unavoidable. This is further defined as "recurring, not caused solely by indirect effects and that occur over the long term in unspecified locations." Advanced conservation practices were the only previous excuse (reason) for programmatic takes. This is now rescinded, as is the five year suggestion, which is extended to thirty years. The rule clearly states that this is for the permitting of wind turbines and no other energy source.[170]

This ignites a slow moving storm that is initially ignored. Like Sandy, the tropical storm forming over the central Atlantic at the same time as the final deliberations, this storm slowly intensifies. The Service struggles with the unintended consequences of their suggested solution. The full consequences are released upon an unsuspecting public exactly 20 months later, on Friday the 13th, December, 2013. As Sandy does her damage to the Eastern seaboard, the USFWS, DOJ, and hapless NGO participants have just begun their awful legacy of destruction.

Some of the legal and regulatory struggle is open to the public. Much is in secret. "The Service's strategy is to implement a defensible process for moving forward with eagle take permitting for wind facilities." The words *defensible* and *take* define the pathway the Service is choosing: legal kills. This desired outcome— legally acceptable kills—defines the Administration's direction. Politics trumps science, law, and fifty years of regulations.

USFWS establishes the Eagle Management Team (EMT) which quickly states the conservation benefits of eagle takes. It then moves towards *programmatic takes*—recurring, multiple incidents. By 2011, an Advanced Notice of Public Rulemaking is set forth for public review and comment. This rule proposal is for 30 year takes of eagles by wind farms, and only wind farms, without mitigation or fear of legal compensation judgment.

Eagle take avoidance is obvious by its absence from the discussion. In fact, the Service believes it is erroneous to conclude that avoidance is the first and best step. Minimization, mitigation, and compensation are unimportant as well.[171] Conservation methods are unneeded. This conclusion is reached via a phantasmagorical analysis of the pit-bull of environmental law, the Endangered Species Act (ESA). As the ESA provides for a No Surprises rule for developers who can be certified, then the same rule should apply to wind farms. Since golden and bald eagles are less endangered than the species on the ESA, their deaths should be held to a lesser standard. Conservation is unnecessary as the birds are doing just fine. Justification for ignoring the two laws on the books regarding purposeful programmatic eagle deaths is in hand. Ignore all else in the name of the politics of climate change.

Mitigation techniques that should be considered include these non-related responses to eagle deaths: lead ammunition prohibition, accidental drowning, vehicular collisions, reduced poaching, trapping and accidental takes, enhanced rehabilitation, and removal of old turbines. Each and all of these should be considered as the only—the only—mitigation methods appropriate for wind turbine kills.

Absent too is any reference to the financial responsibility of takes. These impracticalities are to be avoided, as developers already have significant financial hurdles to the creation of renewable energy sources. Their burden is heavy; allow them to pass through the law. Birds will be saved in the future.

Permitable takes need only be reasonable. If they are, then developers have no further financial responsibility. No additional mitigation response should be required, in particular, no financial mitigation response. No fines. No clean-up cost. No replacement cost. No pre-siting review cost. No cost whatsoever.

Finally, if compensatory damage is to be assigned, it should only be assigned based upon actual body count. Risk is removed entirely from turbine farm siting. It is left to the eagles to figure a way around proposed wind turbine blades. If they can, as a sole example from Alaska appears to demonstrate, then the only risk assessment tool required is for an actual kill inventory. Only known, recorded kills should be used to identify eagle risk. If there are no bodies, they must have worked out a pathway around the blades. If they have, there is no risk of bird-blade interface. The risk of take is to be replaced with the actual measurable take. Imagine that the car you drive kills thousands. As long as it doesn't kill you, there is no risk.

By July, 2012, the public comment period is closed. Now the real power play begins.

How do we know of these secretive follow up meetings? A recent Freedom of Information Act (FOIA) disclosure requested by the American Bird Conservancy (ABC) reveals[172] the meetings of the Group of 16. These meetings were designed to address the challenges facing the wind industry on wind turbine bird kills across America. After the public comment period had closed in July of 2012, these Star Chamber proceedings (behind closed doors meetings) ensued. More than 120 environmental, conservation, Amerindian, and wildlife groups had registered their concerns during the *public comments* phase. Many concerned citizens had registered their worries as well.

ABC had already sent a letter specifically suggesting a halt to these private discussions.[173] The *Federal Advisory Committee Act* sets the guidelines and standards for such meetings. Interior could have simply followed these clearly defined rules. It is even called *the negotiated rulemaking process*! Public transparency would have been the natural result. Rather, Interior held the politburo meetings in secret, by invitation only, with a select group of wind energy industry and environmental groups. It is clear from the significant communications that the latter acted simply as cover for the real decision. USFWS even acknowledges receipt of communication from ABC expressing concern regarding compliance with the FACA.

The weather in DC had been diffident during the entire year of 2012. Tornados had threatened the capital district for several days in the early summer. A dozen touched down in Maryland in June. July was the hottest month on record. The first meeting of the Group of 16 was requested in late August, a normally hot and humid period for DC. The gang of 16 requested (demanded?) an informal meeting, without public participation or transparency.

On October 5, Steven Black, legal counsel to Interior Secretary Ken Salazar, responds to the gang's letter by inviting some other groups. Requests to participate from ABC and various Indian tribes are refused or ignored. Two key California environmental officials are invited: Karen Douglas, California Energy Commissioner, and Michael Pickering, senior political advisor to Governor Jerry Brown. Other interested parties are sent an invite from USFWS.

Interior is equidistant from the Lincoln Memorial and the White House, on C St. NW, between 18th and 19th. It is across the street from Constitution Hall, just down the block from the Federal Reserve. This is the raging heart of the federal beast. The weather is threatening on Wednesday, October 24, as Hurricane Sandy begins its unexpected drive up the East Coast.

USFWS opens a two day session with wind industry and conservation stake-holders on that Wednesday afternoon. While the meeting goes on, USFWS's internal memos note that it has already decided on the 30 year ruling and has little interest in the gang of 16. Further meetings "may not be collaborative."

Sandy passes offshore on the 29th. In DC, five inches of rain and 55 mph winds blow—the top end of the wind turbine scales. The federal government shuts down on Monday and Tuesday, the 29th and 30th. On November 13th FWS is informed that the gang of 16 is encouraging extralegal "eagle conservation plans in lieu of permits." On November 15th, some people lower down the food chain of USFWS want an open forum for the next meeting. Senior leaders in the Service refuse such a meeting the very next day.

David Cottingham, Senior Advisor to the USFWS Director, speaks at the next collaborative meeting on November 29th, hosted by the wind industry. Those present—the *stakeholders*—need to find a way to "enable wind projects to get permitted." The industry states at the same meeting that it is interested in allowing wind projects to kill eagles under research permits. Does this remind you of the whale kill permits or dolphin kill permits the Japanese engender for themselves at each year's International Whaling Conference? The following day Mr. Cottingham reports to the Director that the industry wants the 30 year permits with indeter-minate check-ins.

Two months later, on January 30, 2013, John Anderson, Director of Siting Policy of the AWEA, emails Steve Black at Interior. His email outlines the changes AWEA would like to see to eagle takes. Emails for February 6th, 10th, and 11th are *redacted* regarding the scope, intent, and summaries of the one hour collaborative meeting on the 11th. Redaction only makes the fog deeper, more opaque. One simply wonders, "What's beneath all that black ink?" Congress belatedly calls for an investigation into these redacted pages on March 25, 2014.

To its credit, the Service acknowledges the following:

Eagles are vulnerable to blade strike mortality at wind turbines.

The increased growth in the number of wind turbine projects was of concern.

The Eagle Conservation Plan Guidance could lead to the Service issuing a programmatic eagle take permit.

The final meeting is held on March 27th. It is clearly stated that USFWS "has pointed towards renewable energy development as the underlying impetus for this rulemaking…to meet the challenge of climate change."

The National Audubon Society, NRDC, Sierra Club, Defenders of Wildlife, and the Wilderness Society further state that "there is an urgent need for an overarching national eagle preservation conservation management plan. There are deficiencies in monitoring data collection (current state of bird kills) that must be rectified." They urge no extension of the 5 year take permit and no change from unavoidable to *programmatic* in the take permit language.

The American Bird Conservancy argues that the new guidelines—30 years vs. five—are voluntary, not mandatory, and that the Fish and Wildlife Service was relying "almost exclusively on self-reporting by for-profit companies to tell them whether or not they've killed threatened or endangered species." (*Los Angeles Times* 12/13/13)

AWEA argues strongly in favor of programmatic, authorizing takes that cannot be avoided or minimized. The take permits should be less burdensome. In addition, it argues for a *no net loss* rule to be applied to the *national population* rather than to each *local group*, as has been the case. This opens the door to intrusive kills in one area, such as Texas or California, as long as the national population figures remain fairly constant. It suggests a review of the entire *eagle preservation standard*. It urges financial and legal limits to any *mitigation standards*. These standards should be applied only against fully verifiable eagle deaths. No comments are offered regarding verification processes, although the Service notes that the data is skimpy at best.

The USFWS is entirely reliant upon the industry to inform it about kills.

We (USFWS) lack a comprehensive estimate of avian or wildlife fatalities at wind projects.

We (USFWS) anecdotally receive fatality reports from a few operating wind projects.

Isn't it interesting that professors Loss and Marra are able to find information on takes from more than 94% of the wind turbines in the nation, yet FWS can only rely upon anecdotal information from the industry? Their research is concurrent with these cabals.

No prize for guessing which side won the discussion and influenced FWS in its final determination ruling.

By mid-April, the Eagle Conservation Plan Guidance is released by USFWS. In December, the experimental advanced conservation practices are announced to the public.[174] Interior does a final blackout. The media release is on Friday afternoon, the 13th, at 4 p.m. The assurance of little or no coverage over the weekend buries the news. Only the vigilant, such as the American Bird Conservancy, are even aware, are even paying attention.

The deed is done.

Let's offer the Service a final opportunity to deconstruct their death threats. This is what the Eagle Take Rule says:

Eagle Conservation Plan Guidance (ECPG)[175]

Here are some of the basic points of the ECPG:

Programmatic unavoidable takes are voluntarily permitted.

Wind turbine sites are specifically and solely referenced.

Take thresholds are greater than zero for bald eagles.

Takes for golden eagles are offset by compensatory mitigation.

Advanced Conservation Practices must be implemented.

There are no known ACPs.

Experimental ACPs will be authorized, if warranted.

Cost caps for developers will be jointly established.

Maximum costs shall be proportional to overall risk.

Risk shall be determined by actual body count.

Maximum compensatory mitigation shall be specified.

ECPG will fit within the WEG: Wind Energy Guidelines.

A volunteer, staged approach to an Eagle Conservation Plan will be implemented.

Stage 1: landscape based analysis
Stage 2: site surveys, ideally two years prior to development
Stage 3: model determinate fatality rate estimated
Stage 4: apply potential ACPs; rerun fatality models
Stage 5: monitor fatalities during site operations

The manner in which these stages are to be executed are specified or suggested. Engagement between the Service and developers is encouraged and anticipated. The applied mathematics is detailed in extremis: electrocutions are stated as .0036 eagles/pole/year once relative mitigation productivity has been calculated. Discounts rates have been correctly applied to the equation. Nothing happening here, just move along, move right along.

Look, we are trying very hard here. Give us 30 years to figure this all out. We are all in this together. Just follow our lead. The industry has told us what they want to ensure success. We have agreed. Let's just move on. We are **Saving the Planet**.

Community discussions are avoided. NGO comments have been noted. Everyone is on the same page here: *Climate change demands we take these steps to protect all species from imminent destruction.*

Broken promises.

Laws as well as promises have been broken. They are broken every day on hundreds of wind farms across the land.

A broken Service has its back forced to the wall by the rule breaking vigilantes on the climate change posse. String up the eagles!

This *rare exception* for one industry allowed many to remonstrate against the government's ability to enforce the law. The controversy inside the Interior Department enraged much of the USFWS—in the West particularly.

David Newstead is finishing his doctoral thesis for Texas A&M University. He has been battling windmills for years. From his office overlooking Laguna Madre on South Padre Island, Texas, he can see a set of turbines turning in the offshore breeze of an early March afternoon. Just as Don Quixote, he has failed—so far. "Wind turbines are actually pretty shameful. They kill birds. There is virtually no monitoring of takes. There is no methodology for the little monitoring of avian mortality that we can see. The wind industry feels no compunction to report avian mortality, much less to mitigate the deaths."

95% of the land in Texas is privately owned. Across the entire tier of the western U.S., the figure is quite different: 40% of land is owned by the federal government. As a result in Texas, at least, land conservation holds far more water than environmental activism. The Texas Wildlife Association is composed of large landowners such as the King Ranch. Other NGOs—the Audubon Society, the American Bird Conservancy, the Bat Conservancy, even the Sierra Club—coalesce around this center of gravity in decision making. Yet time and again, these coalitions frame a problem, identify compromise solutions acceptable to all parties, and are quashed by the federal government. FWS, BLM, even the Army Corps of Engineers seem to take direction contrary to the suggestions from these middle ground conservationists.

The wind industry in Texas is large and powerful. Florida Power & Light (FPL), E.ON, and BP Alternative Energy are the players. They make book with the Service. Their demands for long term power generation contracts are coupled to transmission line requirements, as only makes sense. Landowners across which these line run then receive lease income from land use, often for decades. This is at least as much as can be earned from hardscrabble land run sparsely with cattle. These leases then carve up the opposition from the very landowners who try to make compromise with the conservation constituency. It is a tough battle with roughly drawn lines of engagement. As on any battlefield, the corpses of both armies lie along the marshlands of south Texas. These are usually both domestic and migratory fowl.

Let's examine two species, the red head duck and the while tail hawk. These water fowl live a simple life. They eat the rhizome from the salty marshes of sea grass lining South Padre island's Laguna Atascosa Wildlife Refuge. They fly to the fresh water ponds to drink. The wind turbine farms are in the ducks' daily pathway. They fly elsewhere, get dehydrated, or are killed by the turbines as they try to pass through the maze. The result is a complexity of impact on a local and migratory avian population. It ends in functional habitat loss. The absence of ducks encourages developers to apply for permits to build new condos along the beautiful south Texas shoreline. Roads are carved to bring in construction supplies and human habitat, further eroding the delicate marshland. Eateries take the place of sceneries.

The Refuge is piecemealed into oblivion. Death by a thousand cuts. All in the name of advancing human service while reducing human impact. The unintended consequence is species destruction.

The white tail hawk (http://www.allaboutbirds.org/guide/White-tailed_Hawk/sounds) is a raptor. It hunts rabbits, lizards, rats, snakes, frogs, and other birds. Laying one to three eggs each season, it is not migratory. Declaring its territory, the hawk owns the sky without dominating it. With its four foot wingspan and mottled white underbelly, it is easy to identify. A recent study of resident bird kills in Atascosa indicated a figure of two kills each year.

The figures allow wind developers to move on as an organization. Yet, what is the science behind this count? The methodology was scanty. The survey was sporadic. The kills were recorded at varying times of day. The search was less than rigorous.

The result? Fewer noted takes than certainly may have occurred. As the Buteo is not endangered, the losses were considered insignificant—not enough to warrant a full environmental impact survey of the area for all bird deaths. More turbines are being built. Deaths at the top of a food chain are aggregated together with passerine and water fowl deaths. Time of day, season, weather, locale, reporting area, frequencies—none subscribe to a common methodology. The counts are less scientific than sclerotic.

In the search for excuses, science becomes road kill. Migratory patterns in the Gulf can be either across, trans-Gulf, or circum-Gulf (along the perimeter). Impact sites along these flyways destroy birds and result in habitat erosion. The development of uninhabited nesting sites becomes the final stage in the avian lives of so many bird species. When developers suggest active response to turbine/avian interface (bird deaths), it is often *operational curtailment*. We will use radar to tell us when birds are entering turbine farms. We will turn down or turn off the turbines directly in the birds' flight paths. Statistically, this can show benefit to avian populations

Bart Ballard of Texas A&M in Kingsville, Texas has experimented with radar detection of flying creatures. This has been touted as a way to reduce avian mortality by the industry. It has been the focus of several so-called studies by industry focus groups. Several problems ensue:

> Insect droves are mistaken for birds.
> No distinction can be made between bird species.
> Radar counts can be confused with land events.
> Time intervals can be critical to observed behavior.
> Seasons and weather change flight patterns.
> Sighting and intensity of radar varies.

The result is observed flight behaviors that may be passerines, insects or raptors, during day or night, clear and foul weather. No discernable difference can be identified. When do you shut down the turbines? How often do you respond? What is the capital impact? What is the power delivery impact? When do the turbine operators say, "Enough is enough"?

USFWS, the Service, has stated that it has no known way to identify, modify, or placate the problem of avian mortality. The wind industry has offered several suggestions, without seriously implementing any of them—for good reason. The environmental community is frustrated with the lack of response to an ever increasing problem of human/avian interface. The response from the Service is turning a blind eye to the results, encouraging more wind farms to be built, endangerment of more species, and actual encouragement of kills in the name of science.

"Does the Service do this for the electric utility industry or other industries?" Kevin Kritz, a government wildlife biologist wrote in September 2011. "Other industries will want to be judged on a similar standard." Experts working for the Service in California and Nevada wrote that the new federal guidelines will be viewed by the industry, regulators, and the public as written by and for the wind industry corporations, since they "accommodate the renewable energy industry's proposals, without due accountability."

What of our Amerindian citizen cohorts? They have a cultural, historical, social, and religious vested interest in the Eagle Tribes. The USFWS Repository exists for two reasons: to hold evidence of eagle kills as potential criminal evidence and to provide eagles and their feathers to our brethren of the Plains.

On the Wind River Indian Reservation, the Northern Arapaho are preparing for spring, listening for the first clap of thunder. "That thunder represents the eagle hollering. When that happens, that's when everything is waking up. The grass is coming back up, the birds are coming back, the plants and animals that were in hibernation are coming out. It's a new beginning. We say 'ho'hou'—thank you." notes tribal elder Nelson White.

http://www.youtube.com/watch?v=5wwvPLPntZk

Amerindian groups have had to apply for permits to *take* bald eagles as part of their religious ceremonies. They have been denied these permits almost entirely. Only once, in 2012, was the Arapaho Tribe issued a two take permit for their religious ceremonies.[176] The species was removed from the endangered species list in

2007, but remains under the aegis of the Bald and Golden Eagle Act. The take was the result of an internecine battle between the tribe and USFWS via federal prosecutors. They took the case all the way to the Supreme Court in trying to prevent any take, but the Court refused to hear the case, citing religious freedoms. The feds maintained that any bird parts required for religious purposes can be obtained, by permit, from a repository of dead birds maintained by the federal government. The tribe maintains that the parts are rotten and unfit for use in tribal ceremonies. Imagine having to request a *Bible* or *Koran* or *Torah* from a locked library by having to petition the librarian for a copy. Your tax dollars at work.

In October 2012, the U.S. Justice Department reviewed and issued a new ruling on the federal policy regarding Amerindian use, transportation, and possession of bald eagle feathers and body parts. Solely exercising its own discretion, Justice ruled on certain actions and activities by tribal members. All other actions remain subject to federal prosecution. The ruling applies only to federally recognized Peoples. This happens to be less than 5% of the U.S. Amerindian population.[177] The fight goes on for the remaining 95% of the population.

Responding to the 30-year permits to kill issued by the Department of the Interior to the wind industry, the National Congress of American Indians (NCAL), the oldest and most prestigious organization of Amerindians, passed a resolution after meeting with legal advisors of the current Administration. The resolution accuses the Administration of failing to consult meaningfully with tribes while pursuing a rule to lengthen eagle take permits for wind farms from five years to 30 years. The resolution declared that eagle permits should not be issued without the consent of affected tribes.[178]

Add these as broken promises to the Nations. The People have two centuries of experience of promises from the Great White Father in Washington. The Nations have two centuries of experience dealing with the goons in DC. The current Bureau of Indian Affairs (BIA) holds land in trust for the Tribes, yet refuses to allow any development of mineral or hydrocarbon resources. Such development would alleviate the 50% unemployment rates, recidivism, and alcoholism. Poverty is rampant on Tribal Lands, yet BIA refuses to allow any self-determination efforts. Radical environmentalists are the naysayers, once again. It seems humanity would be better placed on the planet as primitives. Deindustrialization is the new mantra of death.

The Nations are the invisible warriors at the forefront of this battle. They are being systemically destroyed via BIA and its NGO cohorts. This is racism by another name.

The stuff that dreams are made of...

The Altamont Pass is the deadliest wind turbine avian killing field in America. It lies in Karen Douglas' and Michael Pickering's home state less than two hours away from their offices. You will recall they are two of the senior California state officials encouraging wind turbine sitings in avian flyways. He was fine with one or two condor deaths each year in the sole remaining California condor flock. That would be the flock USFWS spent tens of millions on and the center which Jesse Grantham ordered so well.

The Altamont wind turbines killed an average of 70 golden eagles each month in 2011 according to the *Los Angeles Times*.[179] A 2008 study found 2,400 raptor deaths at Altamont as well as that of 7,500 other birds.[180] With more than 7,400 towers in operation, Altamont is the largest wind farm in the world. It is also one of the oldest, began in the 1970s, under fears of global cooling, as you may recall. It is San Quentin's death row for our obviously guilty avian criminals.

The U.S. Fish and Wildlife Service (USFWS) clearly states that 440,000 birds were killed by wind turbines in 2011, while 573,000 were decimated in 2012.[181] Fines are stipulated. *Takes* are permitted under certain conditions.[182] This is the license to kill. James Bond would be ashamed to pull this trigger.

The newest take conditions were added in December of 2013. The previous five-year exemption on fines has been changed to a 30-year exemption from fines and imprisonment. Why? The industry needs a tighter assurance of the complete absence of political risk from their financial equations for 30 years. It must be a very tight glove to fit over such an iron fist. Despite all of the tax benefits and guaranteed markets, they still need 30-year political guarantees. Do they know something about their financials that they are not sharing? Is a profit so hard to squeeze out over 30 years that they insist upon removing all risk from their secret equations?

What of fines? The remedial actions required under the previous regulations have received a capital spending limit. All remedial action—reporting of deaths, recording of incidents, remedial actions, any reporting—is voluntary. Revocation of an issued permit is a last resort method that the Service would take, as stated in the regulation.[183]

Applications for the five-year permits have been available to wind energy companies since 2009. Only 50 developers had applied for the permit to kill prior to December 2013. Why? Despite the voluntary reporting, if no permit is in place, no action need be taken to protect avian slaughter. The firms are not required to take steps to reduce kills or to reduce any impact turbines may have on local bird populations. The Silence of the Turbines.

Until very recently, a 30-year take was allowed. "Instead of balancing the need for conservation and renewable energy, Interior wrote the wind industry a blank check," said Audubon CEO David Yarnold in a statement. The wind energy industry has said the change mirrors permits already in place for endangered species, which are more at risk than bald and golden eagles.

The regulation was not subject to a full environmental review. Why? The current Administration classified it as an administrative change. "The federal government didn't study the impacts of this rule change even though the (law) requires it," said Kelly Fuller of the American Bird Conservancy. "Instead, the feds have decided to break the law and use eagles as lab rats."

"It's basically guaranteeing a black box for 30 years, and they're saying 'trust us for oversight.' This is not the path forward," said Katie Umekubo, a renewable energy attorney with the Natural Resources Defense Council and a former lawyer for the Fish and Wildlife Service. In private meetings with government and industry leaders in recent months, environmental groups have argued that the 30-year permit needed an in-depth environmental review. No such review was necessary, as the Administration simply issued the ruling without commentary.[184]

On August 15, 2015, the American Bird Conservancy (ABC) did finally win a court battle against the 30-year take. The U.S. District Court, Northern District of California, in San Jose has ruled that the Department of the Interior violated federal laws when it created a final regulation allowing wind energy and some other companies to obtain 30-year permits to kill protected bald and golden eagles without prosecution by the federal government. The court wrote:

> *"Substantial questions are raised as to whether the Final 30-Year Rule may have a significant adverse effect on bald and golden eagle populations." In particular, the courts cited a lack of compliance with the National Environmental Protection Act (NEPA).*

> *"FWS has failed to show an adequate basis in the record for deciding not to prepare an Environmental Impact Statement (EIS)—much less an Environmental Assessment (EA)—prior to increasing the maximum duration for programmatic eagle take permits by six-fold.*

"While promoting renewable energy projects may well be a worthy goal," the ruling continued, *"it is no substitute for the [agency's] obligations to comply with NEPA and to conduct a studied review and response to concerns about the environmental implications of major agency action."*[185]

Most conservation groups are far more aligned with the wind industry than with ABC. In the South, they call a dog that won't hunt *an egg sucking dog* because it would rather steal a chicken's egg. These non-profit organizations meekly deplore the decisions by the Interior Department and its USFWS. While they readily take your annual contribution, they do little to protect the Public Trust.

The Administration said the longer permit time of 30 years was needed to "facilitate responsible development of renewable energy" while "continuing to protect eagles." Everyone is on the same page here. Renewable energy development has primary focus. A few dead falcons and songbirds are to be studied. Future discussion will be reported as necessary. It is interesting that so many organizations have a lobbyist in DC for their own policy protection and promotion, yet Everyman has no one in DC as a lobbyist. Except perhaps for a business woman from California: http://quixoteslaststand.com/2015/08/12/wow-u-s-republican-carly-fiorina-gets-it-about-wind-turbines.

Irrespective of your politics, she states the issue clearly and concisely. Wind turbines kill millions of birds every year. The response should be that **innovation is far more valuable than regulation**. Businesses thrive on innovation; politics thrives on regulation. This book is not about politics. It is about birds. It is about reasonable and efficient energy sourcing. It is about wise science. Let's take the politics out of the story and get the science back to work.

Meanwhile, in NGO-land, The Audubon Society, National Resources Defense Council, Sierra Club, Defenders of Wildlife, and even ABC are each challenged to balance their anti-carbon polemics with their real concerns for wildlife conservation. They appear to have taken their lessons from the Service. Here are a few quotes from their hallowed halls:

NRDC: *"Addressing issues as complex and important as this one naturally creates some tensions."*

Audubon Society: *"We're between a rock and a hard place. We support a move away from fossil fuels, but there's no question that this issue has strained* *our relationship with the wind industry."*

American Bird Conservancy: *"Environmental organizations support the wind industry, but that doesn't mean you have carte blanche to run roughshod over environmental law."*

Defenders of Wildlife: *"We are trying to get developers to think about wildlife while minimizing adverse consequences."*

The internal responses of those present at these secret meetings in DC during the approach of hurricane Sandy but not quoted here may be read in the FOIA. You have to dig through 800+ pages of redacted information to glean useful information, but it is there. The NGOs see now how they were the front men in the power politics game. DC and AWEA knew the result before these meetings began.

The grifters were the industrialists. AWEA wrote the script. USFWS, DOI, and DOJ approved the lines. Each played the NGOs for their support. Whether you call them patsies, shills, or suckers, the draft agenda was written in 2009. The power play went down in 2012. Hurricane Sandy blew through town just as AWEA and FWS blew through their late fall meetings with the NGOs. The damage was extensive. Everyone decried the results. Monopole towers continue to spread across the American landscape. Birds die. Energy supplies are more costly. Electricity prices rise. Conventional power production hides in the shadows, covering alt-en's many faults.

What we have is more broken promises—this time to our avian associates in the nation's skies.

The first eagle protection law was enacted in 1940. It and subsequent cousins in laws and regulations have now been emasculated by an industry dictating its wishes to a supplicant bureaucracy, a bureaucracy hell bent on protecting us from ourselves. And on making a tidy profit for its vested members. This is the new millionaires and billionaires club.

The climate has indeed changed. Conservation has fallen beneath the brutal axes of trolls seeking to destroy the environment so to protect it.

Some must die so others may live. "Some are more equal than others."

CHAPTER THIRTEEN

OWLS

http://ibc.lynxeds.com/video/great-horned-owl-
bubo-virginianus/nest-brooding-2-nestlings

http://ibc.lynxeds.com/op_video/98806/embed

The wise old owl is a killing beast. She hunts at night. A few are crepuscular or twilight hunters. Her vision is suited to the nightlife. It is binocular, as her eyes face forward, unlike virtually all other birds. Her hearing is her weapon of choice. Feathers focus distant sounds to vector in on the rustling of leaves by small mammals in dense undergrowth of forests and prairies. Her feathers are serrated, thick, and fluffy to nearly completely absorb the sound of her wings. She flies slowly relative to other raptors. She is a slow but silent killer.

Owls are found across the globe. They range in size from six to thirty inches. The largest, such as the great grey owl, have a five-foot wingspan. Fish owls can attain wingspans of 6½ feet, making them the rival of eagles in dimension. She makes her nest in tree cavities and broods one to a dozen eggs depending upon species. The young leave their nests after two to three months gestation, again depending upon species.

Some species are classified as threatened, particularly in Southeast Asia, where they are regarded as a delicacy. Turbines are particularly rapacious given the slow cruising speeds of owls. Night flights during spring and fall are particularly dangerous for these wise avian friends.

The sound of a wind turbine generating electricity is likely to be the same level as noise from a flowing stream 100 meters away or the sound of leaves rustling in a gentle breeze.

Human Health Concerns

Noise nuisance, shadow flicker, aviation interference, EMR, and wind turbine syndrome (WTS) are on the short list of possible human health concerns. Let's review each individually.

Noise nuisance occurs at both high and low ranges of human sound perception. At the high range, noise pollution is a result of the interaction of the blades with wind and mechanical vibration from the equipment, generators, gearboxes, etc. It is a function of the blade tip speed, the rpm of the blade, and wind speed. Noise is measured in decibels with the decibel rating being logarithmic: 60 dBA is ten times 50 dBA. The sound level increases nearly fourfold with a halving of the distance from the source—the blade tip closest to the ground.

The *whoomph* word used in our opening pages is an approximation of the sound from a cluster of wind turbines. Their sounds interact in a disturbing periodicity. Dutch, German, and British researchers have documented the sound, particularly the night effect, which enhances its impression. People living in proximity may, or may not, be offended or annoyed.[186]

http://www.youtube.com/watch?v=wYnNQoTcsHY

As with avian deaths, there are no national standards for noise nuisance. Some states do specify noise levels for time of day and locale relative to urban or suburban communities. Site designers do try to create three color iso-decibel line sound models of turbines for each site, using methods such as ISO-9613-2 with software such as WindPRO or WindFarmer.[187] Solutions suggested include smaller blades, lower rpms, and direct drive turbines.

Low range noise pollution is more controversial. The AWEA suggests there is no evidence of direct adverse human effect.[188] Dr. Nina Pierpont, MD, PhD, would disagree. Dr. Pierpont is the author of *Wind Turbine Syndrome*,[189] in which she discusses in detail her research of and treatment for the disease. You may judge which source has the advantage of a factual record.

http://www.youtube.com/watch?v=KoVKP0G_f8M

Her simple solutions involve distancing turbines 1.25 to 2 miles away from homes. Stories from Europe and New Zealand amplify her credibility.

In the U.K., this story is but one of hundreds in the press: One person who knows this better than most is Jane Davis, a retired health visitor and midwife who lived on a farm half a mile from the Deeping St Nicholas wind farm in Lincolnshire. The eight turbines were built in 2006 and within three days of their becoming operational the Davis family noticed a constant hum emanating from them.

"We had issues with various loud noises and low-frequency sounds that created a hum in the house all the time, not just when the turbines were turning," says Jane. Within weeks they developed a long list of grave health problems. Jane's father-in-law John suffered a heart attack and developed tinnitus, hearing loss, vertigo, and depression.

Mother-in-law Eileen suffered pneumonia and kidney and bladder problems and husband Julian developed pneumonia, depression, and an increased heart rate. All of them suffered from sleep deprivation. None of them had any significant health problems before.

Jane Davis and her family sued Fenland Windfarms Ltd. for noise nuisance in a five-year legal battle she describes as "worse than fighting cancer."

"We finally got to court last summer. We had three weeks in the High Court and I was eight days on oath and five in the dock," says Jane. "The case was adjourned so more noise monitoring could be carried out. But the day before the noise evidence was due to be heard, on November 29, 2012, the case was settled out of court."

This is all Jane can say. In January of this year the family's house was purchased by Fenland Windfarms Ltd for £125,000, 20% below the valuation given by estate agents. It remains uninhabited.[190]

In New York, Mary Kay Barton reports in September 2013,

"As a NYS-certified health educator myself, I was taught that the World Health Organization (WHO) is the Gold Standard when it comes to health recommendations. WHO recommends night-time ambient noise levels no higher than 42 dbA for a good night's sleep. The sound being emitted by industrial wind factories (in upstate New York) is much higher than that. As I told you, a friend of mine had the sound measured INSIDE his home to be above 70 dbA."

Her work on wind energy in upstate and western New York is detailed and well documented. 56 towns in New York now have restrictive ordinances on the wind industry. This hardscrabble lawsuit has been a headache for Albany since 2012.[191] The author of the cited piece, Robert Bryce, is a Fellow at The Manhattan Institute. This is a conservative think tank, so many readers may discount his point of view. Most will be open to valid information from all sources.

Let's go to the ground, in Michigan. Cary Shineldecker has a lovely home in rural Mason County. He and his wife have no political axe to grind, no ideological story to sell. They do have serial and serious health problems related to siting of wind turbines around their home. The closest turbine is 1,200 feet away, 500 feet closer than the wind farm's manufacturer states as a safe distance. They have wind turbines out of every window. His health has deteriorated so much that he sleeps on an airbed to avoid the vibrations from the throbbing machines.

http://www.youtube.com/watch?v=3Q4cJ0m821g

http://www.youtube.com/watch?v=lm00e8J6qT8

Both the AWEA and the CWEA have argued against audible concerns regarding wind turbines and their siting in rural residential areas. Their paper makes sense, of course, in that it regards audible sound levels. The challenge is less about audible noise than about infrasound, below 20 Hz, or sound waves below the frequencies that our ears are designed to accept. Their commentary ignores these concerns. Low Frequency Noise (LFN) is a cause of vibro-acoustic disease (VAD). It is common to pilots, flight attendants, and residents of wind farms. These are not related to WTS. These sounds can be as disruptive to ordinary life for those who experience the subtle throbbing.

Sound is also a local perception issue. Some communities may find the sound of turbine blades moving the air a refreshing, futuristic ambient background noise. Others may find it a complete nuisance. Thus, each community should be able to respond to the blade noise, measured or perceived, differently.

http://www.youtube.com/watch?v=PEnL7meWzBc

Shadow flicker is a result of blade rotation at sunrise and sunset. It impacts people and structures within 1,000 feet of the turbine. Germany is the only nation with specific guidelines for flicker. They allow no more than 30 hours annually of shadow flicker and 30 minutes each day. Those suffering from epilepsy are most at risk from shadow flicker, at frequencies above 10 Hz. The industry claims a range of .6 to 3Hz for today's turbines.[192]

Read Chapter Six of John Etherington's *The Wind Farm Scam* and enter the Downton Abbey of cloistered Scotland and North England. These people don't take kindly or quickly to strangers from London and the pages reek with stories of verisimilitude from developers and the regulators. Sound familiar?

Of particular interest to our British friends is the impact of wind farm installation on their extensive peat cover in the north of England and in Wales, Scotland, and Ireland. This is a low population area, one which has the best chance for ideal wind conditions—it faces the North Atlantic. Recall our early discussion on pressure differences over land and water. As a storm roils in from the Atlantic, its low pressure can be amplified by the land. Beaufort Force 9 wind strength is at the top range of wind farm energy production. As long as it doesn't exceed 56 mph, the farms are operating at peak capacity. Just so designed.

Wind Turbine Syndrome was identified as a health concern in the epony-mous book by Dr. Nina Pierpont, MD, PhD.[193] She describes her own voyage of discovery into this new disease. It affects people who live in proximity to wind farm towers. The medical complaints are: migraine, motion sickness, vertigo, anxiety as well as noise, visual, and gastrointestinal sensitivity.

Dr. Pierpont describes these as a "coherent and interconnected neurologic complex." She highlights the first studies as discussed in 2007 by the National Research Council (NRC).[194] A subsequent investigation in New Zealand from March 2009 brings to light further detailed evidence in support of human health concerns from wind turbines.

http://www.youtube.com/watch?v=IEh3sooKU8A

A further investigation by Ms. Lilli-Ann Green across twelve nations inter-viewed dozens of people who have been or lived in close proximity to wind farms. Her story is here in a nearly two hour video:

https://www.youtube.com/watch?v=H5j-EftX8U4

A shorter version of this video (only seven minutes) is found here: https://www.youtube.com/watch?v=IEh3sooKU8A. This is more factual regarding physical conditions associated with proximity to turbine farms. Proximity can mean as far away as five miles. Symptoms disappear when the residents leave or when the turbines are quiet. It is clearly stated that this is not scientific research and that improved research is required to learn more. Australia is considering now what to do to distance wind farms from homes, but no action has been enacted by the federal government of Australia, as of yet.

Ms. Green is an active environmentalist who lives in a passive solar home. She is as far from a fanatic as you could discover. Her concern is proximity and health issues. Consequences are dire and adverse. If you don't give a fig about birds or radiation or electricity rates, perhaps human health concerns are important to you. They are important to these 25 families. Listen to their stories. This is a series of anecdotal stories, rather than a scientific process. The science has to follow from these tales of terror. These people are environmental families trying to be organic, carbon neutral, thoughtful global citizens. You don't have to pay $550 for these reports; they are free for you to observe. The simplicity of the presentation belies the sophistication of the story: multi-generational farm families protesting and winning against turbine companies and local governments without extremism or fear.

The latest from a variety of sources indicate a growing health concern among peoples from many nations that have adopted wind turbine energy sourcing.

Inhabitants of the Tararua Ranges near Palmerston, New Zealand, more than two miles away from a wind farm, reported the effects of wind turbine noise as similar to that reported by Dr. Pierpont. In this case, vibration was the effect reported. The paper showed that Rayleigh Waves, a seismic energy wave, were transmitted through the tower foundations into the inhabitants' homes. These waves were felt most vividly when lying down, in aural proximity to the sound source. The investigators concluded that seismic effects should be assessed when siting offset distances for wind towers.[195]

In Finland, a family has to move because of the blades.[196]

Glenmore, Wisconsin, a town of 1,200, has halted the wind industry from construction. They have persuaded their local health board that infrasound is a danger to residents.[197]

In Germany, proof is presented at a recent medical conference on the WTS issue and on infrasound concerns. The paper is the first to clearly identify the ability of the human brain to respond to sounds as low as 20 dBA. These levels can be felt as far away as six miles and are directly responsible for sleep deprivation, tinnitus, dizziness, and heart palpitations. The wind energy capitalists have continually denied such a possibility.[198]

The association of German medical professionals has called for a halt to all further wind turbine installations until further medical research can be carried out to determine appropriate siting limits to suburban homes.[199]

The Vermont Public Service Board has taken an interesting path. While the World Health Organization says that an average sound level of 40 dBA is enough to cause sleep interruption, Vermont authorities have indicated a level of 45 dBA averaged over an hour can be reported as a nuisance and has given residents a number to call to report such noise.[200]

This is a recent video on WTS from our Danish friends, 30% of whom have their health affected by wind farm proximity:

https://www.youtube.com/watch?v=Rm1b11YCwWg

This was initially set by law at four times the height of the tower. Now the government is reviewing studies suggesting a distance of 16 times the height. Both Vestas and Seimens are strongly opposed.

The Danish High Court ruled in September 2014 that families with homes too close to new turbines were right to claim for health and property damages resulting from their proximity.[201] The current result? Turbine farms are buying entire villages and bulldozing them rather than put up with the noise from residents.[202] Fairly drastic action to take as a final solution to the noise abatement problem in the heart of Windland, Denmark. The home of Vestas, the global leader in wind energy turbines, the Danes lay claim to the most energy produced from wind, 25% at times. They are a very fair minded people and have laid out exactly how much compensation should be paid to each resident afflicted by noise and infrasound. It is cheaper by the dozen to simply wipe out the village.

These health stories could go on for many pages. The health of neighbors decreases with nearness to the turbines. The number of people affected increases with the increase in the numbers of towers constructed. This Vermont site lists dozens of current legal and medical concerns.[203] There is common ground to the symptoms, including the proximity to turbines and the elimination of the symptoms upon leaving the area. Correlation is not causation, certainly, but these hundreds of problems should warrant medical evaluation and diagnosis.

The refusal of the industry across the globe to even acknowledge a health issue is astounding. You would think they would actively pursue proofs against the health hypotheses. You would think they efforts to remonstrate against these complaining crack pots would be legion. You would be wrong.

Yet another broken promise.

The promise of good health from a safe energy source has been ground into the dust by global examples. Don't let this fool you. Much health science is based upon the struggle upstream against long held beliefs in placebos and phantasms. Ignoring these issues is far easier than throwing money at them. Buy up the entire village to silence the critics, both real and potential. Fund bogus research demonstrating nothing. This is serious money we are talking about.

Microwave and aviation hazards are another concern. Flight paths near wind farms must take these into account, as must radar used for aviation guidance systems. Your cell phone signal from microwave towers can be affected by wind turbines. Site design must take their presence into consideration, as must future cell tower siting. What is called *the Fresnel zone* must be avoided by wind turbines according to the FAA. No such avoidance requirement exists for avian safety, of course.

Did we mention flight pattern interruptions for small aircraft? Below is a typical image of wind conditions downstream from a wind farm as far as the eye can see. Turbulence has been reported as far as 8 miles away in South Dakota to the FAA.[204]

RAF fliers have expressed their concerns about wind turbines monopoles for several years. 59 near misses include 15 high risk problems regarding turbine sites near airports and approaches. Almost daily manual updates to flight charts are made to display turbines and their airflow interruptions, as seen in the above photograph.[205]

The issue of aesthetics is as subjective a story as you can possibly imagine. Investigate this and you are immediately in the land of opinion, as wide and deep as humanity. Everyone has an opinion. Regulators try to append rules. Citizenry are equally appalled and pleased. The sight of one or a hundred wind turbines marching across the headlands or moors can leave you speechless with beauty or in horror. Leave it to our British cousins to both allow thousands of wind towers to be constructed across their ancient lands and to protest such connivances of corporate/regulatory welfare.

Last, but not least, **ice throw** can be a problem for northern towers. Ice storm build up tends to be in thin sheets along the trailing blade edge. If the turbine is then spun into generation mode, these sheets can be thrown from the blades. Much like ice buildup on trees or power lines, this is to be avoided as dangerous. The throw may not travel far from the tower base, but it can be deadly. Canadian concerns have been repeatedly expressed.

Human health is supposed to benefit from renewable energy. We have seen the deaths from cancer and respiratory disease in China as rare earth minerals are mined for two-ton wind generator magnets. We have seen the current local health concerns expressed by hundreds of residents near wind turbines. The promise of a healthy day from wind energy dissipates in the glare of another noisy day and night for many who suffer. They were not consulted when siting review for their local blade health busters was initiated. They are rarely consulted after the construction. They are often viewed as cranks and eccentrics with hypochondriac tendencies.

These are promises broken by those who assured us of a bright tomorrow.

CHAPTER FOURTEEN
HAWKS

https://vimeo.com/11885574

After a short scuffle, a gloved hand drew out the huge old goshawk, her barred chest feathers puffed into a meringue of aggression. Her feet were gnarled and dusty, her eyes deep, fiery orange. She was beautiful. Like a granite cliff or a thundercloud. She completely filled the vet's room. She had a massive back of sun-bleached grey feathers, was as muscled as a pit bull and intimidating as hell. So wild and spooky and reptilian. Carefully, we fanned her great broad wings as she snaked her neck right around to stare at us, unblinking. The narrow bones of her wings and shoulders (were) light as pipes, hollow, each with cantilevered internal struts of bone like the inside of an airplane wing. We checked her thick, scaled legs and toes and inch long talons. This goshawk was much bigger than me and much older: a dinosaur pulled from the Forest of Dean There was a distinct, pre-historic scent to her feathers, peppery, rusty as storm rain. — Helen Macdonald, H Is for Hawk, Grove Press, 2014, pp 18 - 19

Little more can be said. Ms. Macdonald knows her birds and her words. Both of her books on raptors, this and Falcon are ready to read, exciting and very informative. She writes poetry.

Innovation Encourages Success

What does today's wind turbine technology offer, at the cutting edge? How can these new ideas replace older approaches? Are they cost effective at the economic and environmental levels? Can regulators move sufficiently quickly for adaptation?

All capitalist societies are based upon choice. Companies offer choices in the marketplace. Consumers choose, according to their whims, interest, pocketbook, time of life, time of year, curiosity, drive, personal economics, and a host of other drivers.

Until an outside agency intervenes to change the process. Ancient regimes would dictate the terms of economics. Guilds replaced this dictatorship with a looser confederation based upon the skill of workers. Firms did the same as mercantilism took shape. Unions, cooperatives, soviets, and clubs tried to establish formal rules for manufacturing, production, distribution, and sales. The Great Depression offered an opportunity to globalize the *dictatorship of the proletariat*. In America, the federal government took over from the states.

Each of these market interventions changed the way society exchanged wealth for goods and services. Labyrinthine passageways have replaced the outright corruption of organized states and mafias. Regulations exist to protect us. From whom or what is less important, it would appear. Each of these interventions began with good intent. Each intervention has resulted in a *reductio ad absurdum*—a rule is false if its acceptance leads to an absurd result.

Kill birds to protect them.

Choice has been removed from the decision making. In wind turbines, as we have seen, the choices have been reduced to one. The Franken Tower, a three blade Cyclops with its one massive leg firmly planted in the ground turning in the wind gusts, devolving free power from Gaia to Mankind.

There are other choices. We shall examine them here. They have been ruled superfluous. Regulations dictate; regulators have decided. Uniformity encourages rapid propagation. The distortion of Henry Ford's dictum, "You can have any color car you want, as long as it is black," has drummed up a nightmare of incessant noise and death. Soviet style decision-making is obvious.

This is excessive, according to the wind doctors. Why should you want more? You have had enough. You need fewer choices, wiser ones, to regain your health. The planet is sickly, sickened by the pollution of humankind. We need to reduce the choices and move forward. Think a five year plan for the 21st century. Time is of the essence. We have to intercede or we will all die. We will kill the planet.

I beg to differ. There are choices. As we have decided to accept the foreboding results of the AGW hypothesis, the more choices or ideas that we throw at the problem, the quicker we should be able to resolve it, eh? With these rigors in mind, let us examine the pantheon of wind power generation facilities. Remember that Edison tried hundreds of filaments before he settled on the universal one in the incandescent bulb today. (Yes, even that universal was eliminated by regulatory dictat. These bulbs are now illegal.)

Failure is an assumed characteristic of capitalist behavior, indeed of all animal behavior. A major league baseball hitter with a .300 average is very well paid indeed for failing 70% of his times at bat. The statistic is far less important than the lesson. Failure is hard wired into our genetic makeup. It is how we survive. Reduce the failure rate to zero and you may have a brilliant design, or you may have ignored critical elements. The design will fail across several dimensions. You ignore these failures at your own risk. Wind energy sourcing, engineering, placement, monitoring, transmission have all failed. These failures are now ignored. The risk grows.

Perhaps the most obvious, yet most ignored challenge is an engineering one. Wind energy suffers from the fact that it is not despatchable. Consequently, it may generate electricity when it is not needed and may not generate electricity when it is needed most.

The most important engineering task is to store wind energy for future use. This is also the most technically challenging aspect of power generation—electrical power storage. Without reliable and affordable energy storage systems it is impossible to replace conventional power generation. Reducing their use is not viable. It is uneconomical. It usually creates more carbon dioxide rather than less.

Storage is the real gold ring in alternative power generation. No source will be a true replacement for conventional power from coal, gas, and nuclear until such time as electrical storage is viable. Battery storage works in your car. It can work in your home. It does not work at the industrial or commercial level. It does not work at the grid level.

So, it is the greatest challenge. In engineering, many approaches wisely attempt to deal with the most difficult obstacles first, knowing that realistic answers to simply gigantic problems are an engineering solution away. The really big ones require the most time and effort. Therefore, they should be addressed first.

Today, this challenge is, at best, addressed in an offhand manner. Very little funding goes to electrical power storage, either governmental or private. Every inventor and physicist has a small working model or idea for a model on their workshop bench. No one has a real solution. Stories abound.

Perhaps this is the new frontier in alternative energy solutions. Perhaps some work bench will produce the solution, just as has happened in telecommunications, media, and technology. Alexander Graham Bell, Walt Disney, and Thomas Edison set the stage for each arena. Dozens have followed in their footsteps. The cell phone, video, and laptop toys we play with today were each an idea in an inventor's mind. Each developed without government support. Each may be the leading character for the new future of alternative energy. We shall see.

A final observation. The newest class of capitalists has the moniker: *enviropreneurs*. These are people who seek local answers to local environmental issues. We shall use just one of hundreds of fine examples: the cooking stove.

Cooking has changed dramatically in the U.S. since the late 1940s, when the author was a small child. It was labor intensive, included wood and or coal, was poorly ventilated and costly. Technology has changed this beyond recognition. Yet, in much of the impoverished areas of the wood, the major pollutant is smoke from heating and cooking fires. These fires burn dung, wood, charcoal, coal, and refuse. They do so in poorly ventilated small structures without insulation and overpopulated with humans, vermin, rats, insects, etc. nearly four million people die each year from this air pollution. That is far more than from HIV, TB, and malaria combined.

The Shell Foundation and environmental scientists from Colorado State University designed and produced low cost fuel efficient stoves that burn these traditional biomass materials. The firm, Envirofit, produces *profit for purpose* stoves for dozens of communities around the world.[206] Eleven million tons of CO_2 have been saved. More than 1,000 jobs have been created in home countries. The result is a 60% reduction in fuel use, a 50% improvement in fuel efficiency, and a 50% reduction in cooking time. Women have to spend far less time seeking fuel and cooking meals. Freedom allows pursuit of family and economic pastime.

The simple choice of better design, the easy application across small villages and the ready acceptance by local families makes the benefits clear. And what was the cost? It was a gift from Shell. No strings. No tax credits. No incentivized regulatory dictats. The environment would be the better by 400 million tons of reduced CO_2 if such a simple device were in place around the world. Note that the one original stove has morphed into dozens of choices dependent on local fuels, cooking styles, design choice and consumer ideas. One size has evolved into many. You could buy a few hundred of these stoves, send them to whatever village you chose and see immediate personal, cultural, social, economic and environmental impact. Try it. Contact the website. Go ahead, we'll wait.

You see, Schumpeter is right. Small is beautiful you have just made a small contribution, relative to your wealth. The impact is tremendous and grows each week from your gift. Cost effective actions such as these stoves are an example of the entrepreneurial spirit in action. Let's see how such small enviropreneurial spirit works in the wind energy space.

CHAPTER FIFTEEN
GRIFFONS

http://www.youtube.com/watch?v=D9rMK223D7E

Known by many as a vulture, it is actually a griffon, aka *gyps fulvus*. Weighing 15 to 30 pounds, with wingspans of as much as seven feet, they prowl the skies seeking food from death. Their effortless use of individual feathers to guide their flight path is obvious.

http://www.youtube.com/watch?v=U2CxK-TQxkM

They are huge, like their New World fellows, the condors. Living as long as forty years, they are the trash men, the tipsters for the globe. They rule from mountain crags, the mate giving birth to one chick a season. A *kettle* is formed, a hunting group, as food is sighted. In the trees of the African savanna, the groups are *committees*. Feeding from a carcass, they then become the *wake*.

We have seen vultures, *gyps africanus*, across the African plains, sweeping the skies above the savannahs of Tanzania, the veldt of Namibia, and the Sahel of Mali. The white-backed vulture is closely related to the griffon, its European cousin, and to its white rumped Indian fellows. Eagles, hawks, kestrels, osprey, kites, harriers, owls, and buzzards are far more distant relatives across the Atlantic.

The committees in the trees meet and mate. As dawn warms the chilled evening air, they rise into the sky, one by one, heavy as they loosen their gravity bindings. Once airborne, they soar to a thousand meters or more, mapping the day's conquest. As one bird catches sight of a kill, perhaps by the sound of a lion or the hustle of hyenas, a kettle quickly forms. The ingredients for today's meal are coming together. One by one they land, well away from the feeding frenzy.

Patience is a virtue for these sky masters. Hours may pass. Eventually, one bird approaches closer, then another. Soon they are burrowed deep in the bowels of the recent kill. As the lions rest and clean their bloodied fur with incessant licks and pawing, the vultures feast. They fight with hyenas, foxes, and one another. A thousand birds may feast upon an elephant carcass, hundreds upon a water buffalo or giraffe. The scrum burrows deep into the belly, wrestling bone bits and fleshy morsels. Their efficiency is remarkable. In a day, little will remain to darken the grassy stubble around the kill.

http://www.youtube.com/watch?v=D9rMK223D7E

http://www.youtube.com/watch?v=WxW6t8-1UA8

A 40 pound bird can consume nearly half its weight in meat.

http://www.youtube.com/watch?v=TzmM_8BIx-Y

Once engorged, he will step aside, head back, crop bulging, in a torpid state. His stomach acids can easily breakdown flesh and bone as well as Botulinum toxin, hog cholera, and anthrax.

As the center of gravity shifts lower, he attempts flight. Lugubrious steps across the plain, grasping leaps into the air, failure, then final success as a warm vortex lifts him skyward. He gyres upward, soaring to his nest to feed his mate and her chick.

As he lands, she prostrates herself beneath him, to assure one another of their relationship. They mate for life and depend upon each other for survival. He will disgorge his crop; she eats the larger pieces, while feeding the remaining to the chick. She drops a single egg once a year, incubating it for two months. She remains nest bound for the four months it takes the juvenile to finally fledge, to become airborne. From the nest edge, dropping into the warm air, it instinctively soars upward on outstretched wings with new found confidence. The cycle of scavenger life continues.

Design Alternatives

We shall lead with an extended example of wind turbine alternatives, then follow with two dozen other choices with some interest.

Perched on the mesa guarding the western approach to Madrid and La Mancha to the south, Avila is a walled city. The four meter thick defenses were completed in just eight years and you can easily see the Roman tombstones used in the outer perimeter. The small laboratory of Vortex lies just inside these balustrades. Down a flight of stairs, we meet David Suriol and David Yenes, two of the three engineers who founded the company a few years ago.

They saw opportunity within a disaster. The pulsing Washington bridge was oscillating in the turbulent violence of a windy Sunday afternoon in 1950. Concrete and steel wriggled impossibly, describing a sine wave that soon collapsed into the river below. The engineers saw what the wind wrought: accessible energy.

Their *vibrating stick* has attracted capital from across the globe. Both crowd sourcing and serious capital have funded their initial research. Their international patents protect their vision as it incubates. Their proof of concept was made real in December of 2015, a nine-meter tall pylon that generates 100 KW of electricity.

Distributed technology is the key to the Vortex story. We have looked at their technology and found it simple, as in fewer moving parts. It is environmentally friendlier than any other current turbine technology: the carbon footprint is far smaller, the avian mortality rate will be zero at any scale, and the O&M will be nominal. Costs are projected to be lower at the gross and per unit levels. While potentially less energy efficient than the monopole design, its smaller landscape footprint allows a greater number of units in the same area. Advantage Vortex.

The economics are appealing. Application is anticipated for office buildings, homes, and small villages unconnected to any grid. It pays for itself, when conjoined to a solar facility, in five years. If a stand-alone, the pay back is about seven years, far less than solar today. As a small personal unit, it can power a cell phone or computer in the field. As a mid-sized piece, it can power a cell tower, an oil rig, or any number of smaller commercial systems.

Environmentally, it has no avian impact. The Vortex uses little or no concrete and no steel. While neodymium remains the permanent magnet, the size is scaled down by orders of magnitude, at 1 to 2 kilos per MW vs. 1,000 kilos per MW. With virtually no moving parts, the gallons of lubricating oil required for annual maintenance of the monopole design disappear. Human maintenance and local terrain destruction is nominal.

The grid impact is small both as a power distribution source and as a disruptor. Capital costs are anticipated to be at least an order of magnitude smaller per MW. Human health impact is unknown, but it emits infrasound at less than 20 Hertz—it is thus far more silent. Wind use range compares favorably with current turbines at 4 to 54 mph. The Vortex simply stops working above these wind speeds. There is no need for mechanical gearing to save the unit from self-destruction, as is not the case with today's beasts.

Production should be outsourced to areas of Spain with high unemployment, paying fair wages. 3D printing technology has allowed its developers to design, experiment, and fabricate at a fraction of the cost of a decade ago. They may simply allow end-users to 3D print the masts as needed, while supplying the electronic components at a reasonable price.

The greatest threat may be legal. International patents are fine, but many societies bear little respect for such legal niceties. Control of the application of brilliance can be a challenge for the best of inventors.

Given the small scale nature of their initial marketing approaches, they are creating a platform of opportunity for many, particularly those who are currently disadvantaged by the vertically integrated energy markets of today. Social impact is rarely discussed at the Franken Tower offices. Vortex lives for the concept. Application begins at the home or village level. $800 builds the four meter mast that supplies 100 kW of electricity for a home or small village. No moving parts means nominal maintenance, much like solar. In conjunction with a solar panel of similar power generation, the power source becomes even more applicable to small and mid-range demand. With no need to connect to a grid, it defines distributed technology. Larger masts, or a series of four meter units on a roof top, can support or augment commercial or office building applications.

Those in Germany who have seen a tripling of electricity rates or those in Spain who have seen a doubling of electricity rates welcome the concept. Many homes now have installed solar capacity. Imagine adding a few masts to these individual power sources. No need for a tie-in to the grid, simply consume the electricity as it is produced. Vortex's altruism is welcome in today's world of "what's in it for me?" While striving to earn a decent profit, they are quite willing to share at least a part of their success with anyone owning a 3D printer.

Time will tell, of course. But of the many choices listed here in this brief over-view of monopole alternatives, Vortex offers tremendous value, at the environmental, capital, distribution, and human levels. Let's wish them the best chance for success. Birders, lend you ear. For far less than the cost of solar, you will be able to have a Vortex unit on your property in a few years. You will save birds, the environment, and energy costs. You will do so immediately. You will recoup your capital cost in less time than with solar. You will thumb your nose at power companies and turbine makers. You will save on taxes and reduce regulatory oversight. You will not need a dime of tax credits or false legal structure.

In a few decades this example of distributed technology may change the way in which we produce power—at local, state, and global levels. Much of today's technology is only a decade old, or less. Opportunities continue to abound for much of the global population. With this sort of inexpensive, immediately applicable technology, there may be no need for new grids in the developing world. The developed world may turn away, at long last, from vertically integrated power production and distribution. Legal and regulatory constraints in support of guilded industrial interests may become remnants of an ancient, vertically integrated society. By 2030 power generation may be less costly, more readily available, and universal. Birds and the planet will be co-beneficiaries.

In Complexity Theory and in Chaos Theory, structure evolves from within. It evolves in real time without direction, rather than dictated by vested interests. Vertical integration becomes unnecessary and wasteful. Adam Smith is proven correct, yet again. All are benefited. Costs and CO_2 decline. Health, legal rights, education, and power are evoked from the genius of a few caring engineers. Imagine a world of 7 billion people—each with a college degree, a clean environment, and saved supplies of food and water. What theological, religious, or quasi-scientific theory could hold sway?

If this type of change can happen in medicine, in manufacturing, and in transportation, why not in power production and distribution? Only the power of the wrongfully wealthy can intercede. Only concentrated wealth and power selfishly directed can harm the people and the planet.

The greatest challenge any new technology faces lies in supplanting the current common answer accepted by all. Change is very difficult to accept. Further, when regulations insist upon design commonality, when production costs can be controlled by runway manufacturing, when capital outlays can be apportioned according to a universal blueprint—when the common becomes universal—new ideas are hard pressed to intervene. They are simply interference in progress.

Nice idea. Let us know when you get the funding.

Cost effectiveness, environmental and economic, is both easy to measure and difficult to project. As we have seen in the economics discussion, multiple variables make estimates of future use and success impossible. Complexity theory suggests that the future will remain unpredictable. So, we will avoid guesstimates of the future validity of any approach—fool's gold at best.

Other Wind Turbine Designs

We present here a phantasmagoria of wind turbine design ideas. Many are experimental; a few are operational on a small scale. Breaking through the industrial barrier to competitiveness is a task for the hardiest of entrepreneurs. You have to be crazy.

There is no purposeful order to these choices. There are many other ideas.

Wind turbine design choices can be broken down into HAWT and VAWT: horizontal vs. vertical axis wind turbines. Horizontal wind turbines are the pinwheels you see today. They are ubiquitous because of the nature of the industry's evolution. As verticals have been long scoffed at for their puny size, they are often deployed in small site functions. A brief review of the pros and cons:

	VAWT	HAWT
Lower power, efficiency	x	
Higher power, efficiency		x
Lower heights & production	x	
Greater heights & production		x
Higher rotation speeds	x	
Lower rotation speeds		x
Uneven torque stresses	x	
Absence of yaw device	x	
Ground based nacelle	x	
Absence of wind seeking	x	
Elimination of avian casualties	x	

In reviewing the literature on wind turbines from just a few years ago, a 2009 edition of *Power from the Wind* discusses trends in design and production. One of these future trends is the "greater generator efficiency, in large part by incorporating rare earth permanent magnets."[207] We have seen the environmental results and human health result from this incorporating process.

While greater efficiency and resulting power production from tall monopole structures have their way today, one would think that a field of smaller turbines at lower cost and of simpler design would both have a better return upon capital and be less in need of government subsidy.

For small homes and villages, the **Savonius** appears to be the solution. Simple, cheap, and easily repaired are its key advantages. It has a low power coefficient of 20%. Details here: http://www.windstuffnow.com/main/vawt.htm.

The **Darrieus** is the eggbeater design. You can see one in the background of a recent Tom Cruise film, *Oblivion*. Its 55.4% coefficient is similar to the standard 59.3%. All mechanical equipment is ground based, so there are no massive tower stresses to contend with; this makes for ease of maintenance as well. This is a detailed comparison of the two designs: http://windturbine.webs.com/vawt.htm.

The **Giromill** has a design in line with the Darrieus, with more aeronautically designed blades in the shape of airfoils. Again, all installation and O&M are on the ground, reducing costs. Here is a recent video showcasing these two wind turbine designs: http://www.reuk.co.uk/Giromill-Darrieus-Wind-Turbines.htm.

The following designs are perturbations on or flights of fancy away from the HAWT/VAWT dichotomy.

First is an older list from 2009 of wind turbine alternative generators: http://www.popularmechanics.com/science/energy/solar-wind/4324331. Even five years ago the ideas were useful. Today's approaches, however, seem far more advanced.

Here is the latest technology from the Netherlands: http://www.greendiary.com/bladeless-wind-turbine-for-urban-communities-ewicon.html. The **EWICON** converts 7% of the wind energy into electrical energy. With improvements, the professors at Delft University say they can increase that efficiency to 30%. You will recall the figure of 48% as the norm for three blade turbines.

Elena is an innovative approach to wind turbine design. Check it out here: http://www.windenergy-share.com/index.php/innovative-wind-turbine.

The shrouded turbine is far smaller than today's turbines, while significantly increasing energy efficiencies. They are seeking funding and prototype design.

Wind lens from Japan amplifies wind speed and increases the efficiency of production at the power source. While suited for smaller projects, it may grow to more massive applications. It is quiet, as well. Check out this wind turbine design here: http://www.youtube.com/watch?v=vQexzNg_e9A.

Another very quiet VAWT wind turbine design is showcased in this video: http://www.youtube.com/watch?v=pBWBQYJVbYU. Again, the smaller application today may limit its usefulness from the perspective of governments, but homes and commercial buildings may be able to make significant use of such designs.

Add solar panels in an artistic manner while increasing energy production: http://www.youtube.com/watch?v=LWEK4qRALS8. The choices do appear endless, rather than limited.

Sheerwind (http://sheerwind.com/benefits) offers greater power output at reduced costs from lower wind speeds, while having no impact on wildlife. The absence of moving parts allows the unit to capture the slightest breezes through a tunnel and accelerate wind speed by a factor of four. It is avian friendly, as there are no moving parts. At half the capex and nearly half annual O&M costs, it hopes to recover original outlay in less than five years. Its cost of $.02/kwh would be far lower than even the best dual cycle natural gas turbines.

Gedayc is a Spanish invention. Actually it is the award which this invention won for David Sarria. The artistry may be more interesting than the design functionality: http://www.youtube.com/watch?v=8WQpEkZf8Ks.

SkySails (http://www.skysails.info/english/power) is the most interesting power source, from a sailor's perspective. Large, automated kites are sent aloft to gather energy from the ever blowing wind at higher altitudes. While this design is at a developmental stage today, the promise is for ease of access to greater wind power generation. Here is a video: http://youtube.com/watch?v=LIjBv5bbAM8.

Kitgen is an interesting idea from our childhood. Check it out here: http://www.youtube.com/watch?v=YoYru_BMbpg.

Vortex has a crowd-sourced funding idea with its alternative wind turbine idea. Taking advantage of *vorticity*, this turbine model is driven by naturally occurring wind vortices. It is virtually soundless, has zero impact on avian populations, and with no gearing and mechanics uses very little for lubricants. The co-CEO, David Suriol indicated in *Wired*, June 2015.[208] "It's pretty cool-looking. It looks like asparagus. It's much more natural than giant pinwheels."

In its current prototype, the elongated cone is made from a composite of fiberglass and carbon fiber, which allows the mast to vibrate as much as possible. An increase in mass reduces natural frequency. At the base of the cone are two rings of repelling magnets, which act as a nonelectrical motor. When the cone oscillates one way, the repelling magnets pull it in the other direction, like a slight nudge to boost the mast's movement regardless of wind speed. This kinetic energy is then converted into electricity via an alternator that multiplies the frequency of the mast's oscillation to improve the energy-gathering efficiency.

There are no gears, bolts, permanent magnets or mechanically moving parts, which encourages lower production and operations costs for Vortex. Based on field testing, the Mini ultimately captures 30% less than conventional wind turbines, but that shortcoming is compensated by the fact that you can put double the Vortex turbines into the same space as a propeller turbine.

It's less expensive to manufacture, totally silent, and safer for birds since there are no blades. Vortex Bladeless says its turbine would cost around 51% less than a traditional turbine whose major costs come from the blades and the extensive support systems required. Its carbon footprint is 40% less than that of a typical turbine. Current generated electricity cost is less than 4 cents per KWh, on a par with natural gas. Should you be interested, here is their website: http://vortex-bladeless.com/home.php.

Here is the Dutch university at Delft trying kites on for size: http://www.youtube.com/watch?v=FJmlt3_dOuA.

The absurd is also possible, at least in these tests: http://www.youtube.com/watch?v=hm4Vu-yEswo.

Windtamers claims to be able to take more than the theoretical maximum kinetic energy of 59.5% from the air. Diffuser technology claims the success, although this video is rather old: http://www.youtube.com/watch?v=-YJuFvjtM0s.

Vertical axis wind generators (http://www.gizmag.com/dabiri-fish-school-wind-farms/28355) are the concern of John Dabiri's counter-rotating arrays. Modeled upon schools of fish, the intention is to autonomously generate wind power in all situations. The idea apparently has merit, if not full funding. Distributed power generation vs. centralized generation is the root of this solution, preferring the former, as per the Rocky Mountain Institute's Amory Lovins. Follow Mr. Dabiri here: http://www.youtube.com/watch?v=x2audOlniaQ#t=19. He is a MacArthur Fellow, one of our current kettle of geniuses and a remarkably likeable fellow.

Ogin Energy (http://www.oginenergy.com//sites/default/files/OginFAQ-Frequently-Asked-Questions_0.pdf) uses a shroud, a cowling, which wraps around the extremes of the blades, acting as protection for avian interface. Their physical function is to enhance air flow. They claim "3x the energy output per unit of swept area." Their 100 kW-class machine ranks as mid-sized with lower initial capital as well as lower ongoing O&M costs. Construction and siting issues are greatly reduced, from months to days. Noise and nuisance factors are lower. Proximity to distribution lines reduces or eliminates high voltage connectivity, substations, and long distance transmission lines. The units are applicable to greenfield projects and onsite needs. Wake recovery issues are reduced by half, allowing sites to be more compact.

Back to sailing metaphors that so interest this author, the **Saphonian** rig (http://www.saphonenergy.com/site/en/how-does-it-work.59.html) has design elements from its African origins. A recent TED exhibit, the promise is great, the product still in its infancy. The sail-shaped bladeless design frees birds to return to their wilderness condition without fear of being mauled. Recall the Betz Law that defines the limits of energy extractable from a wind source? This sail claims freedom from such restriction.

Another sailing based VAWT solution is found here with Peter Coye of California Energy and Power: https://www.youtube.com/watch?v=NiHG5ahf4zQ.

Uprise Energy (http://upriseenergy.com) has the compulsory revolutionary name. They offer the Portable Power Center:

> *The Portable Power Center and Energy Conversion System are currently in a developmental stage. The wind energy products feature currently available technology blended in an intelligent manner, utilizing sound engineering principles. All innovations demonstrated fit squarely within the core competencies of Uprise Energy and have undergone independent, third party review by top-tier engineering firm,* Quartus Engineering.

> *"To summarize, it appears that you have identified a combination of technologies that can together enable new levels of performance. The anticipated manufacturing costs you reviewed with us, at volume, do not appear out of line with the proposed technologies. Further, the proposed technologies are functionally proven in similar applications."* — **Quartus Engineering**

It must be said that today's turbines have their share of problems, as you can see here. This collection of videos from around the world give fair testimony of the challenges in operations and maintenance for today's wind farmer (http://www.youtube.com/watch?v=liNIqYNHRXE and https://www.youtube.com/channel/UCNKCRnV1Hj_dWpd9q4B574g).

New technology applies to prevention as well as avoidance techniques. DTBird and DTBat are European. They are in place across the continent and have a site in Wyoming. Their technology gives warning of approaching birds and bats so the operator can turn down, turn off, or flare the blades to avoid dangerous encounters (http://dtbird.com/images/download/DTBat_Datasheet_03.2015.compressed.pdf). This is their collision avoidance model in action: https://www.youtube.com/watch?v=qFF9_ZEt6CU.

The self-appointed assayer of current wind technology is Mr. Barnard: http://barnardonwind.com/2013/06/03/good-and-bad-bets-new-wind-technologies-rated. We are indebted to his website for many of these recent ideas. He offers a public service, free of charge, for those who want to learn all he knows about wind and its energy generation potential.

He is hard core AGW (anthropogenic global warming) and will not tolerate deniers or billionaire fossil fuel companies. That said, he offers his own views on the newest wind technology, ranking each by thirteen different criteria. Certainly his critique has real value.

It must be viewed within its deconstructionist context: nothing compares to the devastation to come from a warming Earth; all means necessary to prevent this holocaust are worthy; the predictions are perfectly and totally accurate; and any dispute with these forgone conclusions is heresy. Viewed through these lenses, one can only see the future as he describes it. Yet, his clear understanding and description of a wide variety of wind turbine alternatives is educational, certainly.

CHAPTER SIXTEEN
FALCONS

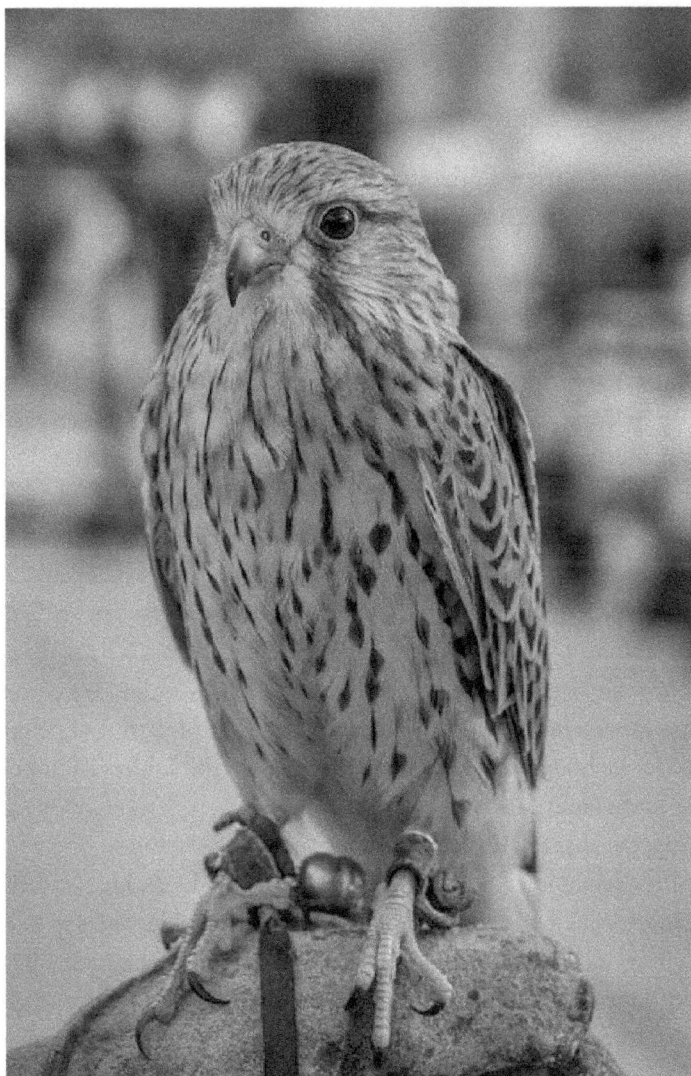

https://www.youtube.com/watch?v=uv_2u64R0eA

https://www.youtube.com/watch?v=pITpkU2w3tM

Any man who thinks he can be happy and prosperous by letting the government take care of him had better take a closer look at the American Indian.
— Henry Ford

What Would You Do?

Your first step may be to go birding. Walk outside to watch and listen to the birds in your neighborhood. They are talkative neighbors, certainly. Entertaining, yes, but they are also teachers. The lessons? Listen. Watch. Be still within to hear their chatter:

https://www.youtube.com/watch?v=X_aDelb_gFg.

That's what happened to Garry George. An associate introduced him to birding. Growing up in Texas, he had always loved the outdoors. He joined field trips at the local chapter of the Audubon Society. Soon he and his partner David were traveling to Papua New Guinea to log sightings of the rarest of birds, the birds of paradise. Of 45 varieties, Garry has seen 29 as of his last count. An avid birder, he has 6,400+ sightings of the 10,000+ extant species on the planet today.

Garry is the Chapter Network Director and Renewable Energy Director for Audubon, California. Tall, lanky, and quietly dressed, he is a serious participant in the avian mortality story at the national level. When he smiles, his corduroy blue eyes crease a face etched in the sun. His knowledge of birds and of people is deep and broad. As such, he brooks no small dealings on the subject of importance to each of us: bird takes.

Gary suggests that NGOs such as Audubon and the American Bird Conservancy (ABC) are the true representatives of the people and the birds. He also feels strongly about alternative energy sourcing. The challenge is to bridge the gap between the perceived energy needs of today's consumers and the impact on nature from energy production. During our interview he offered insight and careful thought on the challenges of birds and blades. While we respectfully agreed to disagree on many issues, his vision is important to mainstream birders as represented by Audubon.

Join the American Bird Conservancy, the most important bird advocacy group in America: http://www.abcbirds.org. The American Bird Conservancy is a group further out on the edge of the envelope in expressing alternative solutions, different approaches. They are less fearful of the USFWS and have won several court battles. Once you join, you may find yourself on a field trip to watch birds. You may begin a log. You may become deeply immersed in the field work and start taking trips to Borneo and Tajikistan. Or Costa Rica and Asheville.

An important program has been developed by ABC: The Bird Smart Wind Energy Campaign which recognizes the challenges to alt-en conversion. It recognizes the significant impact upon avian and bat populations by the Avatar destruction of their habitat and flyways. To assist with Bird-Smart siting decisions, ABC developed a Wind Risk Assessment Map (http://www.abcbirds.org/extra/index_wind.html). The three components of their strategy are intelligent siting, mitigation, and compensation.

Overlaying the ABC Wind Map with the U.S. Geological Survey (http://eerscmap.usgs.gov/windfarm) and Federal Aviation Administration (http://blog.aopa.org/vfr/?p=1252) maps of existing and proposed turbines, respectively, this study showed that there are tens of thousands of turbines already existing in highly sensitive areas for birds and tens of thousands more planned.

Anyone can visit the American Bird Conservancy site and explore: http://www.abcbirds.org/extra/index_wind.html. Here you can see where turbine sites should be forbidden and where they may be considered. This is a simple overlay of the USGS map with ABC's suggestions. Risk assessments are in red and orange. *Intelligent siting* has been done for the industry by one of the keepers of the public trust.

Odd that this was never considered by the wind industry or its cohorts. What a concept! Consider the birds! The USFWS response?[209]

We are currently in the process of evaluating the efficacy and use of the Guidelines and the Service is considering regulatory options.

They have been currently in the process since 2009. It must be a strong current. As Louie says in Casablanca, "I'm shocked, shocked, that there is gambling going on in this casino."

Mitigation means reducing the deaths. While the industry has tried a variety of approaches—some of which work for bats—no serious industry wide solutions have been brought forth. The DOE says they are still studying the matter: "Technologies to minimize impacts at operational facilities for most species are either in early stages of development or simply do not exist."[210]

Compensation would, in their world, take the form of a fee or fine for every turbine in the killing fields. These funds would be used by public trust agencies such as ABC to continue conservation research for endangered species. The funding mechanism is already in place with the Migratory Bird Act and the Endangered Species Act. USFWS would hold and allocate the funds. It has the power now to do so.

We have argued in favor of wise wind power manufacturing, siting, and power sourcing; far greater awareness and prevention of avian mortality; design choice in an open marketplace unfettered by regulatory or preferential intervention; and the elimination of government price supports for production, siting, and distribution.

If you have made it this far, you are aware of the wind power design alternatives. You may have begun your own exploration of these alternative wind technologies. If so, you have quickly discovered that the few selected here are but the tip of the iceberg. Technology is the rampant field of mankind. We are nothing if not inventive. From sticks to fire, from wheels to bricks, then to weapons and welfare, the species *Homo sapiens* is well designed to design. Your inquisitive nature brought you to this book, these ideas. This nature is what will continue to wrought change in a new, yet strangely calcified industry. The bounds of bureaucracy must be lifted.

The author sees that, in negotiations between industry, regulators and NGOs, the public is strangely absent. The Department of Justice's political motivation is clearly more concerned with parsing old laws and treaties to the advantage of the industry than they are at enforcing a decades-old and revered law between states and the native nations. The Department of Interior sincerely caters to its masters at AWEA without reference or ear for the public. They have closed the doors to the public.

In this intransigence of power, perhaps a fairly balanced NGO such as American Bird Conservancy is our best choice. They are broad based, with respect for their years of service to birds and birders. They represent the people in the fields of America. When you join, you allow them to be your eyes and ears. You can do more. Let's hold their feet to the fire of wise and fair decision making. We're certainly not going to get it from the Bolshies at Sierra Club.

Audubon was a participant in the FWS meetings at the heart of this book, while ABC was uninvited for political reasons. The Society was one of the eight NGOs invited to meld policy into a framework of effective regulatory response to the massive increase in wind turbines across America. The concerns expressed by Audubon were serious. They suggested the Advanced Conservation Practices here illustrated:

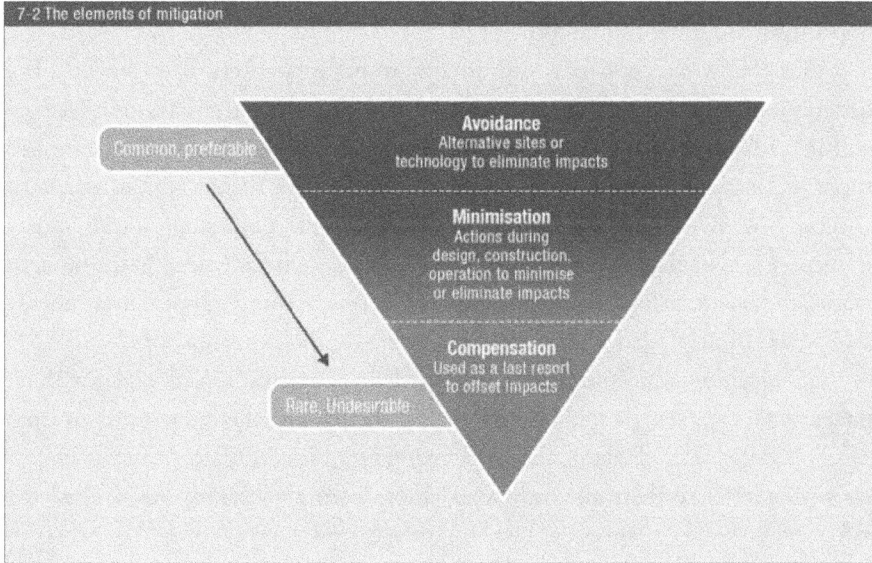

7-2 The elements of mitigation

Common, preferable

Avoidance
Alternative sites or
technology to eliminate impacts

Minimisation
Actions during
design, construction,
operation to minimise
or eliminate impacts

Compensation
Used as a last resort
to offset impacts

Rare, Undesirable

These suggestions are broad and simple. They parallel the precautionary principle. They are completely ignored in the ECGP. Therein, no mention is made of these easy to follow steps. Instead, a five step volunteer, experimental ACP plan is expected to follow soon, perhaps next year; once we get a few ideas on paper.

Apply these simple ideas to the development of any power source. Oil and gas is doing so now, as they create new sites, in conjunction with supportive regulatory state agencies. The wind industry suggests these rules only work on a site by site case. They insist upon their own design of a risk assessment process. Meanwhile, the federal agencies state clearly that this is none of their business; it is all private enterprise working for the public's good. They trust the wind industry. We must do the same. Trust us.

You don't trust the oil and gas industry. Why should you trust the wind industry? Why can't they follow state mandated regulatory guidelines? Why can't they be subject to outside review? Why do they prevent mortality studies from independent scientists? Why do they and the federal government assume that the outcomes of the current studies are confidential corporation information?

If you participate in field trips or in birding, congratulations. You will see the world through the eyes of a hawk, an osprey, a falcon. May these recent steps of discovery in the new world of birding lead you to further field trips. Enjoy the opportunity to learn and tell others.

Your following steps may lead directly to manufacturers, to inventors. The dozen or so radical wind turbine ideas we have discussed here have hundreds of cousins. Like sperm cells, most will die in transit. Yet they will each attempt to succeed. Therein lies the beauty of capitalism. We need no dictation from an august authority. We need innovation, failure, improvement, and ultimately a wide variety of successes. Unicular approaches work well in the unicellular world. In the modern world, they are less than useless; they are a drag on the future. They detract energy and power from myriad solutions that lie just over the event horizon.

Contact your state and federal legislative representatives. Visit them with an image of a locally slaughtered bird if you dare. Better, give them the org chart from ABC on mitigation. Suggest this as a starting point for regulatory review. Follow up with a letter to them the next week. Blog them a few times. Always use the same image—they are good at recalling images and you may not have access to unlimited first edition Benjamins.

Here are examples of local action taken by small groups seeking answers to their real questions. These are from around the world. There are at least 2,200 anti-wind groups globally.

In a recent example from supposedly Socialist Vermont,[211] the citizens of Irasburg voted 279-9 against new wind turbines on their hills. They know their vote is non-binding on the company. They know that the State and the Feds will impose their will, regardless of their wishes. Yet they voted anyway. The vote was resounding. On October 2, 2015, they made their Patriots' point.

An Ohio community halts expansion of turbines.[212] Greenwich Neighbors United, a group of property owners who joined together to coordinate efforts to voice concerns about a proposed industrial-sized wind turbine park in Huron County, applauded the report of the Energy Mandate Study Committee which recommended that there be no further escalation in the mandates until the impact of federal requirements are better understood.

In the U.K., locals have forced wind developers to back away from their plans for further massive expansion: https://wind-watch.org/news/2015/09/25/turbines-turned-down-on-range-of-adverse-effects.

In Nebraska a group of smaller townships have banded together to fight wind development. This is a state with both massive turbine installations and a significantly declining whooping crane population. You can find the details here: http://columbustelegram.com/banner-press/news/townships-line-up-against-wind-development/article_8be33e8e-4ecd-5d6b-9049-ab7449f18244.html.

In San Jose, the California ABC won a federal case against the 30-year extension from five years of endangered species takes. Read about the court case here: http://abcbirds.org/article/bald-golden-eagles-victorious-court-invalidates-30-year-eagle-take-rule.

You should be interested to note than the vast majority of USFWS field operatives find politics dominates decision making at the Service. Read the details here: https://www.hcn.org/articles/political-science.

In Massachusetts, the infamous offshore turbine towers are being fought onshore: http://www.capecodtimes.com/article/20150923/NEWS/150929774.

In rural Washington state, the Whistling Ridge development faces a growing challenge against its exploitation of the local wind alley: http://www.columbian.com/news/2015/sep/18/whistling-ridge-faces-new-legal-challenge.

In the mid-Atlantic Appalachian Mountains, a farm of 131 turbines in 2006 will grow to 16,000 to 32,000 turbines by 2030. Migratory birds most affected will be bats, raptors, and passerines that fly closer to ridge lines with absence of thermals. Pennsylvania has already seen a 1/3 decline of golden and bald eagles there. Avian flight avoidance and displacement behavior will be costly to species with unknown results. The birds will have to fly further, consume more energy, hunt more frequently, and arrive at their destinations later in the season. Five acres of forest are removed for each turbine in Pennsylvania with a total of 20 acres impacted by roads, lines, and maintenance activities. Forested ridges are the worst possible locales for wind farms, given the declines in bat survival rates. The ridges also happen to be the best sites for wind energy production. Initial industry-based studies shut down after a first study showed enormous impact. Cumulative impact results in an explosion of number and types of endangered species.[213]

Forbes magazine comments on the wind project in Ohio halted for the sake of birds: http://www.forbes.com/sites/williampentland/2014/02/02/

worries-about-bird-deaths-derail-wind-power-project-in-ohio.

Here is how it is done in Europe, specifically in Germany and in France (https://www.youtube.com/watch?v=1mWVgrstvdc). The video is in German, but you can run it in translation, or simply get their story by watching. These are ordinary people with simple lives directly impacted by wind farms on a massive scale. They are concerned about their health, their home values, the health of birds, and the environment. Listen at minutes 29 to 31 for the sweeping whoosh sound. Does it remind you of the films *Avatar* or *War of the Worlds*?

These protests are organized locally. They are specific. They demand answers from the wind firms. They are not political. They are humane, heartfelt expressions of families' concerns. You can do this, too.

In Spain there is a cacophony of protests against the environmental and ecological incursions of their nation by the wind industry. A recent photo from a wind site proves their point:

The people in Spain propose a five point plan with many similarities to our own:[214]

1. Immediate cancellation of all wind projects,
2. Elimination of all central planning models, siting and production,
3. Independent avian impact studies,

4. Respect for all current environmental laws, and

5. Wind power to reflect local needs vs. power company demands.

The European wind industry, represented by the European Wind Energy Association (EWEA), is less full spirited than in the recent past. It is petitioning the European Commission for "protection against potential subsidy cuts."[215] This is in response to a halving of subsidies from $121 billion in 2011 to $65 billion in 2014. Their fear is palpable. There are now fifteen different legal attacks in Spain against cuts in subsidies filed under the International Energy Charter Treaty.

Ready to Take Further Action?

Participate in ABC's action plans. Ask your local, state, and federal representatives the following six questions. Suggest these questions as a round table discussion with local community leaders. If the politicians won't set these meetings up, do them yourself. Offend the powerful!

1. Why do we spend federal tax dollars on the development of wind energy?

Initially, we wanted to spur innovation and reduce installation costs. With larger turbines and monopole towers, the power generation efficiency has improved. Costs per MW of generation have declined. Competition from Chinese manufacturers has driven down pricing, not technology breakthroughs. It is an industry fully capable of self-funding through the capital marketplace.[216]

As wind generation has grown, it has become an operational concern for Independent System Operators. The grids cannot deal with its variability and intermittency.

2. Why is the fundamental problem with wind generation its unreliability and variability?

Wind generation is like salt. In small doses it enhances flavor. More is less good. Massive amounts are destructive. So, too, with wind power generation. In areas where wind capacity is a significant, while still minor, percentage of total power sourcing, the problems with wind are serious. The most difficult problem is rapid ramp-up and ramp-down over very large areas.

3. Why must we follow in the footsteps of the Europeans?

Consumer prices for power have tripled in the last decade. Spain is threatened with bankruptcy because of their all-out acceptance of solar and wind sourcing. In Germany, despite having 30 GW of nameplate capacity, periods of regional calm have lasted from a few minutes to several consecutive hours. These threats to power generation are increasing in frequency and amount. Offshore wind is coming to the fore here in America. It has proven a disaster in the U.K. The recent closure of a massive pending wind farm is the first glimpse from England of a clearer path towards the future. The current British government is withdrawing support for wind farms.

4. Why can wind energy companies sell at a negative cost?

Because of the production tax credit, wind electricity producers quite often are bidding negative prices to sell their production. No responsible operator of a conventional plant can bid negative prices. They are blocked out of much of the electricity market when the wind is blowing. Wind energy also gets preferential access to the grid by regulation. As a result there is a growing crisis in traditional, reliable electricity generation. In Europe, all utilities are on credit watch. In both Europe and Texas it is becoming increasingly difficult to get financing for new conventional power generation. ERCOT in Texas is raising the ceiling price in the spot market to $9,000/MWh. The annual average in Texas is $45/MWh. Why is the ceiling being reset at such a lofty figure? All this simply stimulates further conventional power generation requirements as back up generation. Quite clearly, wind is destabilizing the market.

It is becoming increasingly obvious that the more wind generation there is, the more reliable reserve capacity is required. Thus, utilities have to maintain a complete duplication of reliable generation assets to back up the wind farms. That is physically wasteful, economically untenable, and environmentally disastrous. It happens to be far more costly. Where is the logic of sourcing green energy backed up by hydrocarbon-based energy?

We have recognized that the actual capacity at demand times is a small fraction of the nameplate generation, from 20% to 29%. Furthermore, half of that generation takes place at night when demand is low. Peak demand generation is even less. This is because peak demand often occurs during cloudless high pressure weather events in both summer and winter when temperatures will be extreme and winds will be calm.

5. What are the environmental impacts of wind energy?

Interconnections with Norway, Sweden, France, and the Czech Republic allow rapid power variations to be balanced by fast response hydro and nuclear generation. Their nuclear plants will all close over the next decade. The same is true of wind central: Denmark. In Texas, these variations are met by conventional power generation coal and natural gas plants. Keeping these online as *spinning reserves*, able to respond quickly to wind generation fluctuations, is environmentally costly. More CO2 emissions result, not less.

Chinese miners die to purify rare earth minerals for our wind turbines, our flex fuel, and our electric cars. Their waters and lands are poisoned at ever increasing rates without environmental oversight. One-third of all Chinese drink from the heavily polluted Yellow River.

Health concerns of many who live near wind farms have multiplied in recent years. At least three new health issues (not yet diseases) have been related directly to wind farm mechanics. Siting near farms and homes should follow safe distancing guidelines. Turbine towers surround some family farmhouses. If one were to fall, death would certainly result. The responsibility for these sitings is every bit as important as wellhead emissions from fracked oil and gas wells. These environmental concerns for wind farms are entirely absent from the discussion.

Caring about our fellow humans doesn't seem to be a guiding principle when it comes to wind sourced alt-en.

6. Why are these regulatory, systems, engineering, physics, and human health issues the crux of the wind energy problem?

"We have met the enemy and he is us," as Pogo has said. Real solutions are neither top down regulatory nor bottom up market-based. These worldly solutions will instead be multi-faceted, coming from many sources without coordination. The worst applications of command and control, such as we have today, is destroying the best of what is available in wind power generation. Large system solutions with mega-GW beasts of burden crash the very systems they are intended to help. Taxes and regulations have created a Frankenstein monster. It is growing. It is destroying. When we have major grid crashes stimulated by intermittency and irregularity, there will not be a simple pill to solve the problem. We will have gastritis of epic proportions.

On the following pages is a list of further personal suggestions. You can begin in your own home. You can participate in sharing information. These are ways to improve the abuses of the industry while clearly protecting the environment. Spread these ideas to your friends, family, and associates. Tell local, state, and federal politicians. We can all learn from one another if we share.

Begin in your own home.

If important to you, monitor your own carbon dioxide levels. There are various websites that will accept your money in exchange for their indulgence. They offer and promise to pay for a carbon offset according to their particular formula. You may be paying into a pig farm waste recycling effort in Nigeria or a waste water plant in Kyrgyzstan.

Conserve energy.

The single greatest impact you can possibly have on yourself, your family, and your community is an increased awareness of power usage. You can do this in your home, your car, your work, and your lifestyle. But these are simply the beginning. As you consider these, then experiment with them, you will find other energy saving possibilities occurring to you. Try them as well.

If you want to save the planet, begin with yourself. You will find your living costs declining. You will find your attitudes changing. You will find your personal energy expanding. There is no downside to these ideas. Apply what you find of value. Your lifestyle will dictate your choices as will the region of the country in which you live and the time of year.

> Your indirect benefit will be lowered energy consumption. This will lead to lower avian mortality and lower energy bills.
>
> Energy saving tips in your home: http://eartheasy.com/

live_water_saving.htm and http://www.energyideas.org/documents/
factsheets/hometips.pdf.

Energy saving and fuel saving tips for your car: http://www.cars.com/
go/advice/Story.jsp?section=fuel&subject=fuelTips&story=mpgSave.

Energy saving tips at your work: http://www.ecomonitor.
com/12-energy-saving-tips-for-your-workplace.

Lifestyle energy conservation ideas: http://www.youthkiawaaz.
com/2011/05/7-tips-for-energy-conservation-in-our-daily-life.

An Action List

Here are some ideas to promote on wind energy. If you subscribe to the theory
of Anthropogenic Global Warming and if you have made it this far in the book,
then you can have a direct impact upon CO2 emissions with these ideas.

Avian Mortality

- Start birding. Learn from the skywalkers. Spend time with birds on the internet and in the field at sanctuaries.

- Harmonize the various lists and registries regarding definitions of *endangered*, *threatened*, *recovered*, and *extinct*.

- Encourage the scientific approach to species protection, applied nation-wide, affected locally.

- Monitor migratory and breeding events and force wind power feathering and shut downs at peak transition times.

- Siting of wind farms must be preceded by at least a one-year feasibility study and supplemented with annual avian impact surveys.

Design and Construction

- Monopole wind tower designs are but one choice from many.

- There are feasible solutions to tower design. Encourage the capitalist exploitation of their use.

- Raise the standard on federal construction projects to a net benefit level, leaving any affected local species better off.

- Grid efficiencies, including power storage, are the greatest challenges. Face these first.

- Performance-based contracts, regulations, and revenue bargains should be *de rigor* at every level of intervention.

- Integrate all power sources into the grid smartly. Deny no source access via false pricing or regulatory fiat.

Regulatory and Political Action

- Remove tax incentives for production of overly large beasts of burden – smaller works quite often, without tax regimens.

- Tie regulatory impulse to local demand rather than a political animus that changes with the electoral season.

- Rely upon market forces more than regulatory and legal stricture.

Adam Smith knows far more than Mr. Smith in Washington.

If you are concerned about climate change, then acknowledge that the climate always changes. Our first responsibility lies in our actions. Act as you find appropriate for your level of concerns. You may also acknowledge that change is unknowable and ungovernable; we just aren't that smart. Don't give credit to hustlers and don't give money to swindlers. Respect solid research. We should all have an open mind.

These ideas are based upon a simple concept: community. Open and public discussions between all members of society, not just industry and government, is essential to open channel communication as well as ensuring trust in federal decision making. With an open channel that delivers and receives messages, more parties can agree upon more approaches, approaches that may at first appear contraindicated. Without this channel or with a one-way funnel, discord and distraction lead to animosities and disenfranchisement. While that may be the goal of the powerful, it is not the goal of Good Government.

This Is for the Birds:

- Support pre-construction and ongoing site assessment for avian and bat impact.

- Support species-level impact reporting at the local level, paying close attention to high impact species: Raptors, bats, and passerines.

- Monitor insectivore population impacts resulting from loss of predator populations.

Changes are underway as you read this. Local landowners, farmers, and non-militant workers are fighting a grass roots effort to alter the wind energy terrain. In many cases they are winning. Small victories lead to greater opportunities.

The U.S. Court of Appeals has recently overturned the USFWS's extreme application of the MBTA of 1918. USFWS has selectively enforced avian mortality issues by choice of venue and culprit. In Texas, they voted unanimously that accidental avian mortality was not covered by the Act. It regarded intentional kills only.[217]

A local Wisconsin board of health recently ruled that Wind Turbine Syndrome is real and affects local farmers. The appeal was brought by the residents of a small town against further turbine incursions and was approved.[218]

In Maine, a state previously green in its energy sourcing but now addicted to wind, the local Fort Fairfield Town Council approved setback rules far in excess of state rules (written by the wind industry). Locals were fed up with factory closings. Each time all available natural gas supplied power was used by homes, the back up power for wind turbines was cut out of the power grid. Factories suffered closings without electricity. The town had had enough. They said no more wind farms.[219]

Ontario, Canada has gone from 10 wind turbines in 2003 to 1,200 in 2013. Health issues began to appear. Farmers who were happy to lease land for extra income are now demanding restrictions upon the industry. Land values drop if they are adjacent to wind farms. Health screenings are beginning to take place. Work has slowed to a crawl on new turbines.[220]

Scotland has rejected plans to add 14 new larger turbines to the Spango Wind farm. Its proximity to a wild refuge for birds was lauded by many local parties. This follows several other local protests against British wind farming over the past three years. Locals are taking it into their own hands to recommend to their councils that these towers must be halted.[221]

In North Carolina, John Droz, retired physicist and wind energy expert, has filed a suit against the state environmental agency for allowing massive new wind farms to be developed with permits that reference older turbine design and construction rather than following the new law, *Wind Permitting 44 of May, 2013*. Having spent so much on previous permitting, the developer suggested it should be grandfathered. An administrative hearing follows.

CONCLUSIONS

He that will not be counseled cannot be helped. – John Clarke, *Paroemiologia*

http://www.youtube.com/watch?v=l-yo9CYv1mM&list=PL3A07E3FD993AD6B8

Let's repeat a few of the words from the Introduction.

Today's wind energy industry is a disaster. It is incestuous. It is owned by and simultaneously owns big government. Regulatory fiat defines business success. Rent seeking is the ideal business model. A few firms dictate to the regulators while taxpayers foot the bill.

This is an aberration of the truth. Birds die. Billions are wasted. People are sickened and die on our farms and in the faraway deserts of Inner Mongolia. Few consumers and fewer environmentalists are aware.

The grid will fail soon. One of the causes will be intermittency of power supply during peak demand time. The other will be the regulatory *force majeure* of closing hundreds of coal and gas fired power plants.

Wind turbines cause significant avian mortality across a wide range of bird genus. Many of these dying birds are endangered species. They are protected under federal law. The accepted mortality figures from USFWS are guesstimates. They are absurdly low. They are based upon formulae rather than carcass counts. Siting issues are entirely ignored in the design process. Towers are placed in direct line with bird flyways. Alternative placement is never considered. Kills are relegated to the end of the line: "We'll get to them eventually. We have to save the planet right now."

Wind is, at best, an ancillary source of power generation for the nation. Intermittency is the largest engineering obstacle. From this challenge, most other issues arise. Face it and solve it. Smaller challenges will then be more accessible to solution.

The industry is entirely supported by the governments of the world. These governments fear falling skies and rising waters. They proffer tax incentives for the construction, maintenance, and sale of power from wind turbines. They remove all risk from the industry. The result is a fattened calf. The result is massive tax fraud perpetrated by the governments of the world upon their own peoples.

Wind turbines are meant to reduce CO2 emissions for the industrialized world. They are meant to replace coal and gas fired power plants. These are falsehoods. Power plants need to cover for their power production intermittency. New turbines mean more coal and gas-fired power plants. These produce more CO2. They produce more of it than they normally would because they have to remain online even as the turbines are effective. Construction of turbine towers require massive amounts of steel, concrete, and hydrocarbon-based plastics.

Wind turbines are directly responsible for absolute and relative human health concerns in China. Cancer and industrial accidents take an unknown toll of human suffering in the wastelands of Inner Mongolia. These lands have been transformed from desert to waste via mining of rare earth minerals for wind turbine permanent magnets. These turbines are sited too close to farms and residences. The confluence of health issues as expressed by these human neighbors and turbine towers is currently coincidental. This is only because most medical researchers refuse to investigate. The suffering is documented.

Alternative choices abound: in siting, construction, avian mortality, design, and human health. Some of these choices are tested; some are simply interesting ideas. Many are worthy of consideration. Sites can be better monitored over a period of years prior to construction. Tower choices can be expanded beyond the monopole three-blade choice. Human habitation should be much further afield from turbine towers.

P.T. Barnum is rollicking in his grave with the current screenplay over alt-en. Snake oil charmers and backroom grifters have nothing on the con artist supreme. He writes the screen play tale of doom and gloom for the planet. All life will perish if we do not abide by the new green Rulers. The alt-en capitalist backsliders direct the pseudo-drama on the media stage, like the grandstanding ringmasters of the Coliseum in ancient Rome. The media cheer them on and the masses consume their bread and circus. Now there are neodymium watches, computers, and energy sources. All hail Caesar! Ridley Scott could not have done a better job of casting and directing this macabre scene from our 21st century gladiator. The games must go on.

Avian deaths and human health problems are completely unacceptable results of wind power generation. **This game is simply wrong**. Alternative solutions abound. Innovation is superior to regulation. The bigger is better mindset has been accepted by the industry and its regulators. Government sponsored and tax supported projects have an assumption of top-down design. They have a guaranteed risk free return. These methods work very poorly in the open marketplace. The result is either the political redesign of the marketplace, as we have seen in the banking and health care industries during the past five years, or a reexamination of the prerequisites for success by an open eyed and open hearted industry. The latter is preferable in an open, democratic, and capitalist society. Sadly it is lacking today. Today we have Games. We kill countless poor Chinese workers[222] and millions of errant birds, all for the Games. We suffer our farming neighbors their slings and arrows of WTS and low frequency sleepless nights.

The complete lack of accountability is astounding. There is not a line in all the legislation and regulations on alt-en, PTCs, RPS, tax credits, and dregs of the vampirish alphabet soup—not a line!—that requires any accounting of carbon dioxide emissions reduction resulting from alt-en implementation. The absence of proof is proof of its absence. As the Hound of the Baskervilles, in not barking, revealed the culprit in the eponymous story, so this absence of proof of CO_2 emissions reduction is proof of the guilty culprit. The industry and its apologists have failed. They are guilty. That there is today no empirical scientific proof of CO_2 reduction from wind energy sourcing is completely ignored by all participants. Shame on us. We don't even ask for a peek under the tent after paying our quarter for the peep show. We simply walk on by.

The hydrocarbon energy industry can offer the example of a real-time working laboratory experiment in these two approaches. Both work under ideal circumstances. Both fail often. The choice we can see in the energy world is: top-down vs. flight-of-birds management. Let's look at the two sides of a vast experiment in a global laboratory.

For decades, the Big Oil companies have spent billions on mega-projects. These are exploiting the largest known oil and gas reserves on the planet. They are doing so with technology and management that is both state of the art and vertically integrated. Every decision crosses many desks on its way to implementation. Billion dollar revenues are expected, even demanded, from multi-billion dollar projects. Massive drilling vessels sink specially designed pipe into the earth many miles below the sea surface. This pipe then winds its way down tens of thousands of feet further into the Earth's mantle, seeking huge oil and natural gas deposits. Decades of planning, design, and construction leads to huge withdrawals of hydrocarbon from the vast, ultimate deeps. The Gulf of Mexico, the Alaskan Slope, the South Atlantic, and the Indian Oceans are today's drilling sites. $700 million ships drill with multi-million dollar tools to reach miles beneath the sea and earth. Billions in revenue, thousands of jobs, and huge taxes are the rewards.

On the other side of the lab, small independent oil and gas companies, often privately owned, sink similar holes into the deep earth of America. Their own capital is at risk. They are exploring and developing the great shale fields of North America. Their budgets are a tiny fraction of the giants across the room. A well may cost $4 to 8 million and pay for itself in less than a year. They use state of the art technology and equipment. They have the intellectual weight of four and five generations of oil families as well as the experience and skills garnered from time on deep well bores. Their provenance is the farmland, ranchland and fields of America. They share the wealth of their extraction with these landowners, the local community, and their shareholders.

They have discovered what everyone always knew was there, the shale oil and gas fields of North Dakota, Pennsylvania, Ohio, Colorado, Arkansas, Oklahoma, Louisiana, and Texas. Small wells, small capital, adequate returns are shared across the community. In return, America has reduced its crude oil imports by 50% and its CO_2 emissions by 500 million+ tons. They have created more than 2 million new jobs, with 5 million more anticipated. Annual tax revenues to all collecting groups were $85 billion in 2014 and will grow to $1 trillion by 2030. Local natural gas and crude oil are the foundation of the American energy profile. The resource will last for decades. The capital cost is nominal. The Big Oil folks have admitted that they just don't get shale oil and gas.

Two experiments, two results. Hopefully these extended descriptions offer a few ideas on our future approaches to wind energy extraction and exploitation. Small can be better. Big works, too, in the right circumstances. Both earn a good living for their developers, communities, and the nation. Both offer insight and opportunity. Using one design to answer all problems is a disaster. Disease and death do not have to be an acceptable risk parameter. Humans and our avian friends must not suffer at the altar of false gods.

There are choices. Both Texas and North Carolina are abolishing Renewable Portfolio Standards (RPS). You recall these tariff walls built to ensure the success of the wind energy? By ensuring the sourcing of energy from wind farms, they guarantee the underlying costs, making each farm a risk-free investment. They further guaranteed the funds for power transmission lines from the wind farms, at $270 per Texan. That is $7 billion. Now state legislators are renewing and refusing these tariffs. More power to them.

The chickens' fear of the fox has opened the door to the chicken coop. Perhaps he will just eat a few birds and leave the rest of us alone.

Perhaps not. Birds die. Promises have been broken. People are getting sick. Taxes are wasted rather than raised. Regulations reduce incentives rather than increase them. Risk is parameterized away in a hobgoblin of fact-based smoke and mirrors. Few new jobs are created. A few big boys rake in all the dough. The taxpayer is stiffed. The fears of an idiot intelligentsia sell us all down the river.

An attempt to control a runaway climate is resulting in more deaths of more species. The unintended consequences of a less than informed cadre of small-minded engineers bought and paid for by an even smaller group of rent-seeking corporate carpet baggers are making the geosphere worse, not better. It is all summed up by "We need to kill today's birds to save tomorrow's birds."

This is nonsense. There are wind turbine engineers today who are playing with solutions to avoid avian deaths. Designs exist now that are less costly and more energy efficient, while neither maiming nor killing our avian friends. Wind will never be a panacea for our current or future energy needs, but it can provide a sustainable ancillary power supply in such disparate places as smaller communities and rural locales.

Done well and done right, wind power works amazingly well. The energy source is free. Its production and distribution have costs. The turbines can last for decades with nominal O&M. The challenges of intermittency can be overcome with good engineering and good science. Capital can be deployed to good advantage.

Government support is unnecessary and counterproductive. It interferes with the natural mechanisms of the marketplace. One size does not fit all; in fact, it destroys all. Government allocation of capital makes the judgments of politicians and bureaucrats more valuable than field engineers and designers. Consumers get short shrift. Capitalists ride the rent wagon. Political capital replaces real capital: corruption, inefficiencies, and inequality result.

Leaving energy development problems to vested interests with ties to politicians and Big Capital doesn't work for oil, natural gas, nuclear, or renewable energy. Assuming that we just have to throw enough money at a problem to solve it demonstrates a complete lack of understanding of how capital works. Levers are far more powerful than sledge hammers. With the former you can move the Earth. With the latter you can only destroy her.

Allow a thousand turbines to spin. Many will fail, more will survive. Experiments from everyone of interest will result in many failures. The few successes become the solutions for the future. Big money chases small ideas away.

Private capital will always seek its own level. Like water, it flows in the most efficient direction. Gravity is universal, pulling water to its natural source. Like water, capital erodes everything in its path. Brilliant capitalism erodes, so to build. It transfers money from less productive areas to those of greater productive capacity. Just as a river moves soil from its bank to deposit it further downstream, so capital is deposited to new ventures. If you cement a river bed in place, you destroy the riverine ecosystem. If you force capital to flow with regulatory concrete and legal binding, you destroy what you tried to protect.

The ideas shared here are clear, hopefully, to you. The overriding thesis is positive. There is a joy in human progress. We have an extraordinary planet on which to thrive. We have the ability to enrich both it and ourselves from our diligence. Each day is the beginning of history. Freedom perseveres over regimentation. Progress strives forward against stasis. Change is the norm. Loving life and hoping for a better world for our children, our heirs, is the driving force behind every parent's every effort. We naturally aspire to the greater good. We welcome challenge, risk, despair, and hopelessness. From these we achieve the simplest and the most sublime. From darkness we emerge into the future.

This author believes in humanity. We prosper as we grow. Greater population brings greater challenges and opportunities. We are not depleting a resource limited Earth. Rather, we are constantly recycling her riches into new designs. Whether it is wheat or rice, a dog or a donkey, a small village or a megalopolis, each brings new wealth. This is far from a world dark, brief, and desperate. This is a world of choice.

A wager was made several years ago between a world renowned scientist and an obscure statistician. Julian Simon bet Paul Erlich that the price for a basket of raw materials (Dr. Erlich could put into the basket whatever raw material he wished) would be lower a decade hence. The well-known author gladly took the bet, knowing he could not loose. It was a sucker bet, a guarantee. On September 28, 1990, Mr. Simon won by a landslide. The story is simple: fear falls before facts in every contest. We humans improve ourselves and our surroundings as we age. The more we do, the better the world becomes.

That view towards the future is espoused herein. We are blessed with abundance rather than cursed with over population. Each new child brings a new mind to the problems of the future. That mind, well nourished, educated, and loved, has every chance of surmounting each of these unknown challenges. Whether he turns a wolf into a domestic dog or dirt into aluminum or wind into electricity, he will surmount the rising hills and mountains of his future.

Human life is not a menace. We are less enemies of one another than competitors between each other. Stagnation is a temporary alteration of natural progress. Wealth offers intelligence, rights, and safety. Poverty is a condition to be improved upon by any and all means necessary. It is an unnatural condition, one easily remedied by the application of micro-nutrients prior to and just after birth, goodly amounts of clean water, sufficient food of good quality, encouragement from an early age to achievement, respect for each family member, an understanding that competition is normal and natural and, finally, an expectation that constant failure leads to great reward.

Each culture views love in its own fashion. While we may lavish gifts and honor upon it, others may relish it privately by example and introspection. However expressed, it is, like a smile, universal. Hatred, denial, control, unfettered regulation, dictation from afar—each of these is unnatural. While some guidance is always appropriate, less is more. **"Yes, we can"** is more powerful by far than **"No, you cannot."**

Eat wisely.
Sleep well.
Love with abandon.

http://www.liveleak.com/view?i=5ff_1313865345

BIBLIOGRAPHY

Websites

www.abcbirds.org
www.awea.org
www.stopthesethings.com
www.wind-watch.org
www.instituteforenergyresearch.org
www.windpowermonthly.com
www.fuelfix.com
www.eoearth.org
www.iberica2000.org
www.quixoteslaststand.com
www.americanenerggyalliance.org
www.wind-watch.org
www.epaw.org

The list of 2,000+ global organizations against wind turbines can be found here: http://www.wind-watch.org.

Birds

Falcon, Helen Macdonald, Reaktion Books, 2006
H Is For Hawk, Helen Macdonald, Grove Press, 2014
Feathers, Thor Hanson, Basic Books, 2011
The Peregrine, J.A. Baker, NYRB, 1967
The Goshawk, T.H. White, NYRB, 1951

Academic

Mortality monitoring Canada: http://www.transalta.com/sites/default/files/

PCFP_Report5_Dec2011.pdf

Wind: http://www.youtube.com/watch?v=-tdSVqOybqs http://www.youtube.com/watch?v=fUI8nPfmTAA

And the ultimate in storm tossed ships: http://www.youtube.com/watch?v=3eTV4UqukhQ

LCOE: The Past and Future Cost of Wind Energy, NREL, 5/13/12. E. Lantz. M. Hand & R. Wiser: http://www.nrel.gov/docs/fy12osti/54526.pdf

Environmental impacts: National Research Council. *Environmental Impacts of Wind-Energy Projects*. Washington, DC: The National Academies Press, 2007

Engineering: *Wind Turbine Operations, Maintenance, Diagnosis, and Repair*, David Rivkin, Jones and Barlett Learning, 2012

General Readership

Energy, Vaclav Smil, Oneworld Publications, 2006

Energy, the Master Resource, R. Bradley Jr., Richard Fulmer; Kendall/Hunt Publishing, 2004

Energy in Nature and Society, Vaclav Smil, MIT, 2008

Energy Victory, R. Zubrin, Prometheus Books, 2009

The False Promise of Green Energy, Morriss, Bogart, Meiners, Dorchak; Cato Institute, 2011

Fossil Fuels Improve the Planet, Alex Epstein, Center for Industrial Progress, 2013

Power from the Wind, Achieving Energy Independence, Dan Chiras, New Society Publishers, 2009

Power Hungry, R. Bryce, Public Affairs, 2011

Power to Save the World, Gwyneth Cravens, Alfred Knopf, 2007

Powering the Future, R. Laughlin, Basic Books, 2011

Traveling the Powerline, Julianne Couch, University of Nebraska Press, 2013

Wind Energy Basics, Paul Gipe, Chelsea green Publishing, 2004

Wind Energy in America, a History, Robert Righter, University of Oklahoma Press, 2008

Wind Energy Engineering, Pramod Jain, McGraw Hill, 2010

The Wind Farm Scam, John Etherington, Stacey International, 2009

Wind Power, Paul Gipe, Chelsea Green Publishing, 2004

Wind Turbine Syndrome, Nina Pierpont, MD, PhD; K-Select Books, 2009

Windfall, Wind Energy in America Today, Robert Righter, University of Oklahoma Press, 2011

Video sources

http://www.arkive.org/eurasian-griffon/gyps-fulvus/video-12a.html

APPENDIX A

Total raw counts of bird species found as fatalities at monopole wind turbines in the contiguous U.S.: 11/2013.

Species	Total count	Number of facilities with fatalities
Horned Lark	384	26
Western Meadowlark	165	17
Red-eyed Vireo	163	21
Golden-crowned Kinglet	161	30
American Kestrel	117	15
Red-tailed Hawk	103	29
Red-winged Blackbird	97	11
Mourning Dove	92	19
European Starling	88	29
Tree Swallow	81	16
Turkey Vulture	76	17
Ring-necked Pheasant	73	16
Magnolia Warbler	52	18
Rock Pigeon	49	22
Ruby-crowned Kinglet	47	30
Gray Partridge	40	12
Blackpoll Warbler	35	9
Savannah Sparrow	33	14
Yellow-rumped Warbler	33	16
Mallard	32	14
Bobolink	27	6
Brewer's Blackbird	26	4
Killdeer	26	8

Townsend's Warbler	26	9
Chukar	25	7
American Coot	24	11
Cedar Waxwing	22	9
Dark-eyed Junco	22	12
Black-throated Blue Warbler	21	8
Common Yellowthroat	21	12
White-crowned Sparrow	21	8
Wilson's Warbler	21	9
Wood Thrush	21	7
Yellow Warbler	21	11
Yellow-billed Cuckoo	19	6
Northern Flicker	18	14
American Pipit	16	3
American Robin	16	9
American Redstart	15	11
Black-throated Green Warbler	15	7
Purple Martin	15	3
Ring-billed Gull	15	2
Yellow-breasted Chat	15	3
Barn Owl	14	5
Ovenbird	14	10
Barn Swallow	13	7
Wilson's Snipe	13	1
Cape May Warbler	12	2
Swainson's Hawk	12	4
Swainson's Thrush	12	6
Tennessee Warbler	12	7
Winter Wren	12	8
Black-and-white Warbler	11	7
Burrowing Owl	11	2
Chestnut-sided Warbler	11	7

Cliff Swallow	11	9
Warbling Vireo	11	6
American Goldfinch	10	7
Black-billed Cuckoo	10	7
Blackburnian Warbler	10	7
Brown Creeper	10	7
Canada Goose	10	7
House Wren	10	6
Orange-crowned Warbler	10	6
Ruffed Grouse	10	6
Short-eared Owl	10	6
Virginia Rail	10	6
Wild Turkey	10	7
Bay-breasted Warbler	9	4
Blue-headed Vireo	9	6
Eastern Kingbird	9	6
House Sparrow	9	6
Laughing Gull	9	1
Lincoln's Sparrow	9	8
American Woodcock	8	4
Brown-headed Cowbird	8	4
Chimney Swift	8	6
Northern Bobwhite	8	2
Northern Harrier	8	5
Rose-breasted Grosbeak	8	5
Vesper Sparrow	8	4
Yellow-bellied Sapsucker	8	6
Great Horned Owl	7	4
Northern Parula	7	4
Rough-legged Hawk	7	4
Veery	7	4
Black-throated Gray Warbler	6	2

Chipping Sparrow	6	4
Ferruginous Hawk	6	5
Hooded Warbler	6	4
Indigo Bunting	6	4
Red-breasted Nuthatch	6	6
Sharp-shinned Hawk	6	6
Sora	6	4
Swamp Sparrow	6	5
White-throated Swift	6	4
Yellow-bellied Flycatcher	6	4
Golden-crowned Sparrow	5	4
Gray Catbird	5	5
Gray-cheeked Thrush	5	2
House Finch	5	5
MacGillivray's Warbler	5	4
Nashville Warbler	5	4
Purple Finch	5	3
Western Grebe	5	4
American Crow	4	4
Black-billed Magpie	4	4
Blue Jay	4	3
Blue-winged Teal	4	3
Brewer's Sparrow	4	3
Broad-winged Hawk	4	3
Common Grackle	4	3
Common Nighthawk	4	3
Field Sparrow	4	2
Great Blue Heron	4	4
Hammond's Flycatcher	4	3
Hermit Thrush	4	4
Northern Mockingbird	4	3
Osprey	4	2

Pacific-slope Flycatcher	4	2
Palm Warbler	4	4
Philadelphia Vireo	4	4
Rock Wren	4	1
Ruby-throated Hummingbird	4	4
Spotted Towhee	4	4
Bank Swallow	3	2
Canada Warbler	3	3
Cooper's Hawk	3	3
Eared Grebe	3	2
Eastern Meadowlark	3	3
Golden Eagle	3	2
Grasshopper Sparrow	3	3
Herring Gull	3	3
Kentucky Warbler	3	3
Least Flycatcher	3	3
Merlin	3	3
Mountain Bluebird	3	2
Pied-billed Grebe	3	2
Song Sparrow	3	3
Tricolored Blackbird	3	1
Upland Sandpiper	3	1
Western Tanager	3	3
White-eyed Vireo	3	3
White-tailed Kite	3	1
Yellow-throated Vireo	3	3
Acadian Flycatcher	2	2
American Tree Sparrow	2	2
Baltimore Oriole	2	2
Black Rail	2	1
Black-headed Grosbeak	2	1
Blue-gray Gnatcatcher	2	2

Blue-winged Warbler	2	2
Cerulean Warbler	2	2
Common Gallinule	2	2
Cordilleran or Pacific-Slope Flycatcher	2	2
Double-crested Cormorant	2	2
Downy Woodpecker	2	2
Dunlin	2	2
Eastern Phoebe	2	2
Eastern Wood-Pewee	2	2
Gadwall	2	2
Green-tailed Towhee	2	1
Lapland Longspur	2	1
Long-billed Curlew	2	2
Long-eared Owl	2	2
Northern Waterthrush	2	2
Peregrine Falcon	2	2
Pine Warbler	2	2
Ruddy Duck	2	2
Scarlet Tanager	2	2
Sedge Wren	2	1
Sharp-tailed Grouse	2	1
Snow Bunting	2	2
Varied Thrush	2	2
Western Kingbird	2	2
Alder Flycatcher	1	1
Ash-throated Flycatcher	1	1
Belted Kingfisher	1	1
Black-crowned Night-heron	1	1
Bufflehead	1	1
Canvasback	1	1
Cassin's Vireo	1	1

Cattle Egret	1	1
Chuck-will's-widow	1	1
Common Merganser	1	1
Common Poorwill	1	1
Common Raven	1	1
Common Redpoll	1	1
Dickcissel	1	1
Eastern Towhee	1	1
Franklin's Gull	1	1
Great Black-backed Gull	1	1
Green Heron	1	1
Gull-billed Tern	1	1
Hairy Woodpecker	1	1
Lark Bunting	1	1
Lark Sparrow	1	1
Le Conte's Sparrow	1	1
Lesser Scaup	1	1
Lewis's Woodpecker	1	1
Loggerhead Shrike	1	1
Northern Pintail	1	1
Prairie Falcon	1	1
Prairie Warbler	1	1
Prothonotary Warbler	1	1
Ring-necked Duck	1	1
Sage Thrasher	1	1
Scissor-tailed Flycatcher	1	1
Short-billed Dowitcher	1	1
Spotted Sandpiper	1	1
Townsend's Solitaire	1	1
Tufted Titmouse	1	1
Vaux's Swift	1	1
Western Wood-Pewee	1	1

White-breasted Nuthatch	1	1
White-throated Sparrow	1	1
White-winged Crossbill	1	1
Wood Duck	1	1
Yellow Rail	1	1
Yellow-throated Warbler	1	1
Unknown bird	205	26
Unknown Passerine	83	22
Unknown small bird	26	2
Unknown sparrow	24	13
Unknown warbler	15	11
Unknown kinglet	9	7
Unknown *Empidonax* spp.	8	4
Unknown flycatcher	8	4
Unknown gull	8	6
Unknown large bird	7	2
Unknown swallow	7	5
Unknown blackbird	5	3
Unknown thrush	4	3
Unknown vireo	4	4
Unknown duck	3	3
Unknown dove	2	1
Unknown grebe	2	2
Unknown shorebird	2	2
Unknown small passerine	2	1
Unknown wren	2	2
Unknown *Accipiter* spp.	1	1
Unknown *Buteo* spp.	1	1
Unknown *Calidris* spp.	1	1
Unknown goose	1	1
Unknown hawk	1	1
Unknown hummingbird	1	1

Unknown meadowlark	1	1
Unknown medium bird	1	1
Unknown non-passerine	1	1
Unknown nuthatch	1	1
Unknown partridge	1	1
Unknown tit	1	1
Unknown waterfowl	1	1
Unknown woodpecker	1	1

ENDNOTES

A note on endnotes: online research opens a vast array of data sources. The author has listed these without regard to their political views. While the majority of the notes are from official sites, a few are from what may be called inflammatory sites. As with every resource, check it for yourself. A conscientious reader will trust but verify. Sourcing and restricting these minority viewpoints does not reflect the intentions of the author or of the book. You should see all points of view, openly, even as you reject some. *Caveat emptor.*

(Endnotes)

1   http://www.telegraph.co.uk/news/earth/environment/10146081/
Twitchers-flocking-to-see-rare-bird-saw-it-killed-by-wind-turbine.html

2   https://www.federalregister.gov/articles/2013/12/09/2013-29088/
eagle-permits-changes-in-the-regulations-governing-eagle-permitting

3   http://www.eoearth.org/view/article/169244

4   Op. cit.

5   "As many teachers, researchers and other experts in the environmental field know, Wikipedia does contain questionable content as well, and it has, therefore earned a reputation as an unreliable reference when in the hands of the uninformed." Op. cit.

6   http://www.huffingtonpost.com/2013/05/14/wind-farms-bird-deaths_n_3270691.html

7   http://www.sciencedirect.com/science/article/pii/S0006320713003522

8   *United States Biological Conservation, 168,* 201-209 DOI: 10.1016/j.

biocon.2013.10.007

9   http://www.eoearth.org/view/article/51cbf1cc7896bb431f6a6cee/
?topic=51cbfc79f702fc2ba8129ed2

10   APWRA, 2004; chapter 3, p. 52

11   http://www.theregister.co.uk/2015/10/11/freeman_dyson_interview

12   http://ei.haas.berkeley.edu/research/abstracts/abstract_wp262.html

13   http://www.countryguardian.net/Green%20Places.htm

14   Eichhorn M., Johst K., Seppelt R., Drechsler M., "Model-based estimation of collision risks of predatory birds with wind turbines," *Ecology and
Society* 17:1, 2012

15   NOAA library http://docs.lib.noaa.gov/rescue/dwm/1888/18880312.djvu

16   A. Betz, *Introduction to the Theory of Flow Machines,* (D. G. Randall,
Trans.), Pergamon Press, 1966

17   Tony Burton, et. al., *Wind Energy Handbook*, John Wiley and Sons, 2001.
For the engineer and scholar, 780pp of technical descriptions on wind
energy.

18   For a fascinating read of the development of the power grid, try *The Grid*,
Phillip Schewe, Joseph Henry Press, 2007

19   A.G. Drachmann, "Heron's Windmill," *Centaurus*, 7, pp. 145-151, 1961

20   Niki Nixon, "Timeline: The history of wind power," *The Guardian.*
Guardian News and Media Limited, 17 October 2008

21   Todd Malmsbury, "Turbines Spinning an Ill Wind," *Boulder Daily Camera*,

Boulder CO, 1A, 8A, 2 September 1984

22 *New York Times*, 27 June 1976

23 G. P. Tennyson, "Potential of Wind Energy Conversion Systems in the Great Plains," in *Proceedings of the Solar and Wind Systems Workshop*, Lincoln, NE, 1983, p. 78

24 Robert W. Righter, *Wind Energy In America, A History*, University of Oklahoma Press, 1996

25 http://www.telegraph.co.uk/comment/11718550/Why-are-greens-so-keen-to-destroy-the-worlds-wildlife.html

26 PURPA

27 Ros Davidson, "Uneasy and Divisive in the Face of Survival," *Windpower Monthly* 4, p. 14-15, February 1988

28 "Capacity Credit of Wind Power: Capacity credit is the measure for firm wind power," *Wind Energy the Facts*, EWEA

29 "Capacity Credit Values of Wind Power," Wind-energy-the-facts.org

30 http://www.eia.gov/forecasts/ieo/table13.cfm

31 Op. cit., *The Grid*

32 http://docs.wind-watch.org/BENTEK-How-Less-Became-More.pdf

33 "Transparent Cost Database," En.openei.org, 20 March 2009. Retrieved 11 January 2013.

34 http://docs.wind-watch.org/wtn09_Bakker-et-al_Seismic-Effect-3-MW-Wind-Turbines.pdf, p. 2

35   Dr. John Etherington, *The Wind Farm Scam*, Stacey International, 2009

36   UCTE (2007a) Final Report – System Disturbance On ,4 November 2006

37   http://www.stanford.edu/group/efmh/jacobson/Articles/I/USStatesWWS.pdf

38   http://thesolutionsproject.org/#page-fifty-states

39   wingspan.co.nz/birds_of_prey_native_new_zealand_swamp_harrier.html

40   Wind Generation Technical Characteristics, http://www.uwig.org/wind_turbine_tech_charac_draft_final.pdf, p. 17

41   Robert Laughlin, *Powering the Future*, Basic Books, p. 46

42   Phillip Schewe, *The Grid*, Joseph Henry Press, Washington, DC, 2007

43   eei.org/ourissues/ElectricityTransmission/Documents/Trans_Project_lowres.pdf

44   *Energy Efficiency and Renewable Energy*, U.S. DOE, "20% Wind Energy by 2030," http://www.nrel.gov/docs/fy08osti/41869.pdf, 2008

45   http://www.dblinvestors.com/wp-content/uploads/2015/03/Pfund-Chhabra-Renewables-Are-Driving-Up-Electricity-Prices-Wait-What.pdf

46   blog.renewableenergyworld.com/ugc/blogs/2015/08/get_the_facts_wind.htm, 2015

47   http://www.fas.org/sgp/crs/misc/R42023.pdf, p. 7

48   Dan Charles, "Renewables Test IQ of Gris," *Science* 324, pp. 172-175,

April 2009

49 http://docs.wind-watch.org/Intermittency-of-UK-Wind-Power-Generation-2013_2014.pdf, 2014

50 http://www1.eere.energy.gov/wind/pdfs/2012_wind_technologies_market_report.pdf, p. vi

51 http://www.naturia.per.sg/buloh/birds/Haliaeetus_leucogaster.htm

52 U.S. Dept. of Energy, *2011 Wind Technologies Market Report*, p. 33, August 2012

53 http://www.renewableenergyworld.com/articles/2015/07/how-wind-tur-bines-are-becoming-photocopiers-and-the-cost-to-investors.html

54 http://www.scientific-alliance.org/sites/default/files/Intermittency%20of%20UK%20Wind%20Power%20Generation%202013_2014.pdf

55 Wiser & Bolinger, *2011 Wind Technologies Market Report*, p. 23, August 2012

56 http://townhall.com/tipsheet/aaronbandler/2015/07/07/study-cost-of-generating-wind-electricity-48-higher-than-previous-esti-mates-n2021792

57 http://www.theregister.co.uk/2015/06/26/gates_renewable_energy_cant_do_the_job_gov_should_switch_green_subsidies_into_rd

58 Op. cit.

59 http://www.eia.gov/analysis/requests/subsidy/pdf/subsidy.pdf, p. 26

60 http://www.eia.gov/forecasts/aeo/pdf/electricity_generation.pdf

61 http://c.ymcdn.com/sites/www.ospe.on.ca/resource/resmgr/DOC_advo-
cacy/
2015_Presentation_Elec_Dilem.pdf

62 http://www.nrel.gov/docs/fy12osti/54526.pdf

63 http://en.openei.org/apps/TCDB. An excellent visual tool that illustrates a
wide variety of energy concepts beyond LCOE.

64 http://www.masterresource.org/2013/10/
google-green-play-375-million-dollars/#more-28123

65 Op. cit., p. viii

66 http://static.googleusercontent.com/external_content/untrusted_dlcp/
www.
google.com/en/us/green/pdfs/renewable-energy.pdf

67 239,200 x 8,760 x .45 = 942,926,400 KWh. 942,926,400 x .o26 =
$24,516,086.40

68 Undocumented footnote from Jon Boone to http://www.masterresource.
org/2012/07/wind-energy-jobs-myth

69 D. Brady Nelson, Heartland.org, 7 October 2015

70 http://www.fas.org/sgp/crs/misc/R42023.pdf, p. 1, footnote2

71 http://www.census.gov/epcd/ec97sic/def/D3511.TXT

72 http://www.masterresource.org/2012/07/wind-energy-jobs-myth

73 http://www.masterresource.org/2012/07/wind-energy-jobs-myth/#sthash.
e6UDcRAw.dpuf

74   http://www.manhattan-institute.org/html/ir_25.htm#.UGs4Fo6L_do

75   http://www.nrel.gov/docs/fy12osti/52739.pdf

76   http://www.juandemariana.org/pdf/090327-employment-public-aid-re-
      newable.pdf, p. 22

77   http://stopthesethings.com/2015/09/30/
      us-wind-power-outfits-curse-el-nino-for-massive-mounting-losses

78   Tania Jefferies, *Daily Mail*, 29 July 2015

79   http://www.wsj.com/articles/
      sunedison-shares-fizzling-promises-a-new-strategy-1444230335

80   http://www.windpowermonthly.com/article/1367688/
      detail-onshore-subsidy-grace-period

81   Adam Smith, *The Wealth of Nations*, 1776

82   http://www.nationalreview.com/article/425303/
      obamas-disastrous-clean-power-plan

83   conversation with author, 12 October 2015

84   Barry Fisher, "The Threat to Wind Energy," *The New York Times*, 26
      October, 1985

85   http://www.project-syndicate.org/commentary/
      wind-power-wasted-subsidies-by-bj-rn-lomborg-2015-10

86   Bjorn Lomborg, "This Child Doesn't Need a Solar Panel," *Wall Street
      Journal*, 22 October 2015

87   http://www.co2science.org/education/reports/models/models.pdf

88   "Hot Air," *The Economist*, p. 78, 5 October 2013

89   Wikipedia

90   Kate Sheppard, "A Green Tinged Stimulus Bill," *Grist*, 12 February 2009

91   http://www.juandemariana.org/pdf/090327-employment-public-aid-re-
      newable.pdf

92   http://redneckusa.wordpress.com/2009/07/20/
      when-wind-power-blows-jobs-will-fall

93   Peter Harrison, "Once Hidden EU Report Reveals Damage from
      Biodiesel," Reuters, 12 April 2010

94   Conversation with author, 12 October 2015, and Mr. Droz's public
      presentations. Losses from a drop in tourism counts for a small portion of
      these figures.

95   http://docs.wind-watch.org/BENTEK-How-Less-Became-More.pdf

96   www.nrel.gov/docs/fy08osti/41869.pdf

97   This is idiotic. That it was funded with your tax dollars is a crime.

98   Paper presented at the 2009 Minor Metals and Rare Earths Conference,
      Beijing, China, 2 September 2009

99   Mark Smith, "Why Rare Earths Matter," interview with Tom Vulcan, 18
      May 2009

100   Op. cit.

101   http://www.dailymail.co.uk/home/moslive/article-1350811/In-China-true-cost-Britains-clean-green-wind-power-experiment-Pollution-disastrous-scale.html

102   http://www.masterresource.org/2012/02/wind-spin and http://fmso.leav-enworth.army.mil/documents/rareearth.pdf and http://www.vetiver.org/ICV4pdfs/BA09.pdf

103   http://nrmca.org/GreenConcrete/CONCRETE%20CO2%20FACT%20SHEET%20JUNE%202008.pdf

104   http://www.fas.org/sgp/crs/misc/R42023.pdf, p. 8, footnote 42

105   http://www.unitconversion.org/volume/cubic-yards-to-ton-registers-con-version.html

106   http://www.huffingtonpost.ca/blair-king/renewable-energy-canada-ra-re-metals_b_7837710.html

107   http://www.iags.org/rareearth0310hurst.pdf

108   http://wattsupwiththat.com/2015/10/15/are-jagdish-shukla-and-the-rico20-guilty-of-racketeering

109   Op. cit.

110   http://www.academia.edu/9785645/A_Comprehensive_Analysis_of_Small-Passerine_Fatalities_from_Collision_with_Turbines_at_Wind_Energy_Facilities

111   NRC, *Environmental Impacts of Wind Energy Projects*, Washington, DC, 2007

112   Internal email, 28 September 2012, David Cottingham, FWS re: meeting of Group of 16

113   http://www.popsci.com/science/article/2010-10/increasing-wind-turbine-turn-speeds-could-help-reduce-bat-deaths-new-study-says

114   http://www.carbonbrief.org/blog/2013/04/wind-farms-and-birds

115   http://climatecrocks.com/2014/04/24/wind-turbines-hardly-rank-as-bird-threat

116   http://www.pennfuture.org/content.aspx?SectionID=372

117   http://www.fws.gov/southwest/es/documents/R2ES/LitCited/LPC_2012/Johnson_and_Erickson_2011.pdf, p. 8

118   http://eastcountymagazine.org/node/14761

119   http://www.eoearth.org/view/article/169244

120   http://www.livescience.com/31995-how-do-wind-turbines-kill-birds.html

121   http://savetheeaglesinternational.org/releases/wind-farms-to-wipe-out-california-condor.html

122   http://www.iberica2000.org/Es/Articulo.asp?Id=4242

123   http://www.iberica2000.org/Es/Articulo.asp?Id=4382

124   http://eastcountymagazine.org/node/14761

125   https://www.masterresource.org/cuisinarts-of-the-air/usfws-special-agents-speaking-truth-to-wind-power-re-shiloh-iv-part-ii

126   http://www.fws.gov/eaglerepository/statistics.php

127   http://www.fws.gov/eaglerepository/factsheets.php

128 ICF International, *Altamont Pass Wind Resource Area Bird Fatality Study, Bird Years 2005–2010*, M87, November 2012

129 http://www.eoearth.org/view/article/169244

130 See Appendix A for a complete list of avian deaths from the 11/13 Smithsonian study

131 "Agreement to Repower Turbines at the Altamont Pass Wind Resource Area," 2 December 2010

132 http://www.contracostatimes.com/breaking-news/ci_27778194/despite-bird-deaths-electric-wind-farm-wins-extension

133 http://www.learner.org/jnorth/tm/crane/WCEPStats_LossesYr.html

134 http://fws.gov/refuge/aransas/science/whooping_crane_activity_report.html

135 http://whoopingcrane.com/wind-farms-and-whooping-cranes

136 The Group includes Acciona North America; Allete; Alternity; BP Renewables; Clipper Wind Energy; CPV Renewable Energy Company, LLC; EnXco; Duke Wind Energy; Horizon Wind Energy; Iberdrola Renewables; Infinity; MAP Royalty; NextEra Energy Resources; Renewable Energy Systems Americas; Terra-Gen; Trade Wind Energy; Element Power; Own Energy; and Wind Capital Group.

137 http://www.windaction.org/posts/36156-potential-whooping-crane-deaths-demand-eis-for-north-dakota-wind-project-groups-say

138 http://www.birdwatchingdaily.com/blog/2013/10/09/icf-co-founder-george-archibald-discusses-cranes-wind-power-hunting-and-a-bird-named-tex

139   http://www.greatplainswindhcp.org/documents/fact_sheet.pdf

140   Joel E. Pagel, Kevin J. Kritz, Brian A. Millsap, Robert K. Murphy, Eric L. Kershner, and Scott Covington, "Bald Eagle and Golden Eagle Mortalities at Wind Energy Facilities in the Contiguous United States," *Journal of Raptor Research* 47 (3), 311-315, 2013

141   http://goldengateaudubon.org/conservation/birds-at-risk/avian-mortality-at-altamont-pass

142   http://www.kcet.org/news/rewire/wind/california-leads-nation-in-wind-turbine-eagle-deaths.html

143   http://saveourseashore.org/?p=1801

144   http://www.ceoe.udel.edu/lewesturbine/documents/acua_quarterlyreport_fall09.pdf

145   NREL, "Bird Risk Behaviors and Fatalities at the Altamont Pass Wind Resource Area: Period of Performance," 2003

146   American Wind Wildlife Institute, "Wind turbine interactions with wildlife and their habitats: a summary of research results and priority questions," 2015

147   http//www.articles.latimes.com/2013/may/10/local/la-me-killing-condors-20130511

148   http://www.kcet.org/news/rewire/AEWP_ROD_App2_BO_051313.pdf

149   https://earthengine.google.org/#timelapse/v=35.04847,-118.30873,10.397,latLng&t=2.66

150   http://www.bioone.org/doi/abs/10.2193/2009-266

151   Op. cit.

152   http://docs.wind-watch.org/Economic-Importance-of-Bats-in-Agriculture.
      pdf

153   http://www.ncbi.nlm.nih.gov/pmc/articles/PMC3700871

154   http://www.ncbi.nlm.nih.gov/pubmed/25118805

155   http://www.ncbi.nlm.nih.gov/pubmed/25267628

156   http://www.batsandwind.org/pdf/Curtailment%20Final%20Report%20
      5-15-10%20v2.pdf

157   http://www.batcon.org/index.php/our-work/regions/usa-canada/
      address-serious-threats/wind-energy

158   http://www.batsandwind.org/pdf/Arnett_Deterrent%202009-2010%20
      Field%20Study%20Final%20Report.pdf

159   http://batsandwind.org/pdf/Fowler_Final_SRRP_Report_
      January_31_2011.pdf

160   http://www.ncbi.nlm.nih.gov/pubmed/22859969

161   http://www.batsandwind.org/pdf/BWEC%20BIBLIOGRAPHY_
      February
      %202014.pdf

162   http://www.sciencedirect.com/science/article/pii/S0006320713003522

163   http://www.nrel.gov/csp/storage.html

164   http://www.kcet.org/news/rewire/solar/concentrating-solar/

milestone-ivanpah-solar-plant-formally-opens.html

165   Julie Cart, Los Angeles Times, p. A28, 2 March 2014

166   http://kcet.org/news/rewire/wildlife/blm-to-investigate-water-bird-deaths-at-solar-plants.html

167   http://www.kcet.org/news/rewire/wildlife/august-was-a-bad-month-for-birds-at-genesis-solar.html

168   Helen Macdonald, *Falcon*, Reaktion Books, 2006

169   http://www.scribd.com/doc/202165532/American-Bird-Conservancy-FOIA-1, http://www.scribd.com/doc/202165648/American-Bird-Conservancy-FOIA-2, http://www.scribd.com/doc/202165784/American-Bird-Conservancy-FOIA-3, http://www.scribd.com/doc/202166235/American-Bird-Conservancy-FOIA-4, http://www.scribd.com/doc/202166370/American-Bird-Conservancy-FOIA-5

170   *Federal Registry*, Vol 77, No. 72, 4/13/13/proposed rules, p. 22279

171   FOIA, p. 23

172   http://www.abcbirds.org/newsandreports/releases/130219a.html

173   Defenders of Wildlife, National Audubon Society, NRDC, National Wildlife Federation, Wilderness Society, Nature Conservancy, Sierra Club; AWEA, Wind Coalition, Alliance for Clean Energy, NY, Interwest Energy Alliance, RENEW New England, CA Wind Energy Ass., Renewable Northwest Project, Wind on the Wires, USFWS, DOI.

174   These are the FOIA accessed documents as provided to the ABC: http://www.scribd.com/doc/202165532/American-Bird-Conservancy-FOIA-1

175   http://www.fws.gov/windenergy/PDF/Eagle%20Conservation%20

Plan%20Guidance-Module%201.pdf

176 http://news.yahoo.com/wyoming-tribe-gets-rare-permit-kill-bald-ea-
     gles-003151146.html

177 http://www.nativenewsnetwork.com/policy-on-tribal-member-use-of-
     eagle-feathers-released.html. See the hundreds of comments from various
     Native Peoples at this site.

178 http://www.windaction.org/
     posts/39372-native-americans-protest-eagle-deaths-at-wind-farms

179 http://articles.latimes.com/2011/jun/06/local/
     la-me-adv-wind-eagles-20110606

180 http://altamontsrc.org/alt_doc/m30_apwra_monitoring_report_exec_sum.
     pdf

181 http://www.partnersinflight.org/pubs/mcallenproc/articles/pif09_
     anthropogenic impacts/manville_pif09.pdf, p. 268

182 http://www.ecfr.gov/cgi-bin/text-idx?c=ecfr&sid=9a2c074a271d17db-
     16c4a0fa4ca3d2ba&tpl=/ecfrbrowse/Title50/50cfr22_main_02.tpl

183 http://www.salon.com/2013/12/06/for_wind_power_us_extends_permit_
     for_eagle_deaths

184 http://www.huffingtonpost.com/2013/05/14/wind-farms-bird-
     deaths_n_3270691.html. Note the source of these quotes.

185 https://www.masterresource.org/cuisinarts-of-the-air/
     bald-golden-eagles-court-victory

186 G.P. Van den Berg, "Effects of the wind profile at night on wind turbine
     sound," *Journal of Sound and Vibration*, 277, 955-970, 2004

187   Pramod Jain, *Wind Energy Engineering*, McGraw-Hill, 2011

188   AWEA, *Sound White Paper*, 11 December 2009

189   Dr. Nina Pierpont, *Wind Turbine Syndrome*, K-Selected Books, 2009

190   http://quixoteslaststand.com/2012/08/10/
      are-wind-turbines-bad-for-your-health-case-studies-say-yes

191   http://www.nationalreview.com/articles/334102/
      backlash-against-big-wind-continues-robert-bryce

192   NRC, Op. cit.

193   Dr. Nina Pierpont, *Wind Turbine Syndrome*, K-Selected Books, 2009

194   NRC, *Environmental Impacts of Wind-Energy Projects*, National Academies
      Press, p. 109, 2007

195   http://docs.wind-watch.org/wtn09_Bakker-et-al_Seismic-Effect-3-MW-
      Wind-Turbines.pdf

196   http://epaw.org/echoes.php?lang=en&article=n387

197   http://midwestenergynews.com/2015/09/17/
      wisconsin-health-hazard-ruling-could-shock-wind-industry

198   https://stopthesethings.files.wordpress.com/2015/07/full_paper_koch-
      v2-1.pdf

199   http://stopthesethings.com/2015/05/21/
      german-medicos-demand-moratorium-on-new-wind-farms

200   http://www.vermontbiz.com/news/september/

brian-dubie-wind-turbine-noise-what-you-cant-hear-can-harm-you

201 International Law Office, Søren Stenderup Jensen, 1 September 2014

202 Philip Tees, *The Copenhagen Post*, 16 January 2015

203 http://swantonwindvt.org/2015/10/20/
wind-turbines-proven-to-be-threat-to-peoples-health

204 http://stopthesethings.
com/2014/05/09/4-killed-as-plane-slams-into-giant-fans-in-south-dakota

205 http://www.express.co.uk/news/uk/606513/
Air-disaster-making-RAF-pilots-60-close-calls-wind-farm

206 http://www.icontact-archive.com/
Rf_N3EbM5YU4KPbzNPzFPMqkbEP60Pds?w=4

207 Dan Chiras, *Power From the Wind*, New Society Publishers, p. 117, 2009

208 http://www.wired.com/2015/05/future-wind-turbines-no-blades

209 FWS 2014

210 DOE EERE 2014

211 http://watchdog.org/240766/irasburg-rejects-wind-turbines

212 SaveOurSkylineOhio; 3 October 2015

213 Kim Van Fleet: Audubon PA. Dennis McNair, UPenn; Johnston Chandler
Robins, USCS Patuxent Wildlife Research Center

214 http://www.gurelur.org/p/en/projects/wind-power.php

215 http://www.windpowermonthly.com/article/1365584/
     renewable-industry-seeks-eu-investment-protection

216 Or, for that matter, any power production?

217 http://www.eenews.net/stories/1060024567

218 http://www.eenews.net/stories/1060024771

219 http://ellsworthamerican.com/opinions/editorials/feel-good-energy-policy

220 http://thechronicleherald.ca/
     opinion/653454-wind-farms-family-farms-not-the-best-of-neighbours

221 http://www.itv.com/news/border/update/2015-09-23/
     rspb-welcomes-rejection-of-wind-farm-plans

222 http://www.bbc.com/future/story/20150402-the-worst-place-on-earth

www.ingramcontent.com/pod-product-compliance
Lightning Source LLC
Chambersburg PA
CBHW020831210326
41598CB00019B/1871

* 9 780983 573142 *